GW01157859

HITLER'S WAR MACHINE

STURMGESCHÜTZE
ARMOURED ASSAULT GUNS

EDITED AND INTRODUCED
BY BOB CARRUTHERS

Pen & Sword
MILITARY

This edition published in 2013 by
Pen & Sword Military
An imprint of
Pen & Sword Books Ltd
47 Church Street
Barnsley
South Yorkshire
S70 2AS

First published in Great Britain in 2012 in digital format by
Coda Books Ltd.

ISBN 978 1 78159 218 2

A CIP catalogue record for this book is
available from the British Library

Printed and bound by CPI Group (UK) Ltd, Croydon, CR0 4YY

Pen & Sword Books Ltd incorporates the Imprints of Pen & Sword Aviation, Pen & Sword Family History, Pen & Sword Maritime, Pen & Sword Military, Pen & Sword Discovery, Pen & Sword Politics, Pen & Sword Atlas, Pen & Sword Archaeology, Wharncliffe Local History, Wharncliffe True Crime, Wharncliffe Transport, Pen & Sword Select, Pen & Sword Military Classics, Leo Cooper, The Praetorian Press, Claymore Press, Remember When, Seaforth Publishing and Frontline Publishing

For a complete list of Pen & Sword titles please contact
PEN & SWORD BOOKS LIMITED
47 Church Street, Barnsley, South Yorkshire, S70 2AS, England
E-mail: enquiries@pen-and-sword.co.uk
Website: www.pen-and-sword.co.uk

CONTENTS

INTRODUCTION

In a career marked by a litany of mistakes, surely Hitler's biggest mistake of all was to drag a reluctant US into World War II. Even after the Japanese attack on Pearl Harbour Hitler still had the option to keep the US out of the war. In a typical act of delusion Hitler, on 11th December 1941, declared war on the largest industrial nation on earth. From that moment onwards the fate of Nazi Germany was sealed. It took some months to awake the sleeping giant, but once the US juggernaut began to roll the end result of World War II was never in question.

While the US was busy assembling its new armies, navies and air forces, the US Intelligence Service was already beginning to gather intelligence on its new enemy. This information was collated and disseminated to the troops who needed it, in the form of two main monthly intelligence bulletins. These were Tactical and Technical Trends which first appeared in June 1942 and the Intelligence Bulletin which began to appear from September 1942 onwards.

The main focus for the US was initially on the war with Japan and a great majority of the early reports are concerned with the war in the Pacific. However, as America began to come up to speed US forces were soon engaged in North Africa, followed by Sicily, Italy and finally Northern Europe. As the war progressed the requirement for good intelligence of German battlefield tactics became more and more important and in consequence there are more and more reports of German fighting techniques available to us. The vast majority of those reports concerned the fighting in Russia and it is those reports which form the bulk of what you are about to read here.

The material for the two US intelligence journals was originally collected from British combat reports, German newspapers, captured German documents, German training manuals and Soviet sources.

As such the quality of much of what was printed was highly variable, some reports are very accurate, while in others, the precision of the information is questionable to say the least, but that's what makes these reports so fascinating. Regardless of the overall accuracy this is a priceless glimpse into how the men in the front lines learned about their enemy, and as such it presents us with a invaluable insight into how the events of the Eastern Front were perceived at the time when they actually unfolded. The reports also provide us with a host of information concerning the minor aspects of the thousands of tactical combats being waged day in and day out which expand our knowledge of the realities of the fighting in Russia.

Thank you for buying this book. I hope you enjoy reading these long forgotten reports as much as I enjoyed discovering them and collating them for you. Other volumes in this series are already in preparation and I hope you will decide to join me in other discoveries as the series develops.

Bob Carruthers

CHAPTER 1
STURMGESCHÜTZE III & IV

The Sturmgeschütz rumbling forward into action is one of the iconic images of World War II. The StuG, as it is often known, is frequently mistaken for a main battle tank, however, this is something of misconception. There were many occasions on which the Sturmgeschütz (or Sturmgeschütze in the plural) performed the same functions as a tank, as mobile artillery, the StuGs were essentially infantry support weapons and were designed for a specific tactical purpose which was very different to that of the tank formations.

Whereas the tanks were intended to force a breakthrough and keep going deep into enemy territory the StuG, on the other hand, was intended to provide close infantry support against enemy field defences using direct-fire from its main gun. As the war wore on the Sturmgeschütze, with considerable success, were also utilised in the role of tank destroyers and mobile artillery so there undoubtedly was some blurring of the roles.

In consequence of their versatility, Sturmgeschütze III and IV mobile assault guns were manufactured, in a large number of variants, from 1939 to 1945. Over 10,000 vehicles were eventually produced by the hard-pressed German armaments industry making the Sturmgeschütz, in all of its variants, the most produced German armoured fighting vehicle of the war.

Although the Sturmgeschütz III was not actually deployed until 1940, a small number of what were essentially prototype machines were deployed for the campaign in the west. During the course of the war the anti-tank qualities of the up-gunned Sturmgeschütz III were soon appreciated, and as a result the Sturmgeschütz III entered a sustained production run and eventually accounted for the longest

Sturmgeschützschütz rolling across the open steppe during the heady days of the advance into the Russian hinterland during June 1941.

armoured fighting vehicle production run in Germany during World War II. Sturmgeschütze saw action on all fronts and, as the situation deteriorated, they were pressed into service for a wide number of ancillary roles.

The most widely produced vehicle of the Sturmgeschütz family was, of course, the now legendary Sturmgeschütz III. It was not a purpose designed vehicle; as it was built around the proven chassis of the Panzer III tank which was adapted to provide a new type of armoured fighting vehicle for a very specific role. The StuG III was intended to provide a mobile, well armoured, but relatively light artillery piece for close infantry support in the assault.

All StuG IIIs were operated by a four man crew which comprised a commander, gunner, loader/radio operator and driver. The smaller crew and lack of a turret were the main disadvantages of the StuG but the low silhouette to an extent compensated.

The Sturmgeschütz III was still under development and was therefore not available in a combat role for the 1939 invasion of Poland. It finally made its operational debut in France in 1940 where it first saw action in the field but only in relatively small numbers. Only

A Sturmgeschütz commander in action Russia 1942.

twenty four machines were actually deployed, but the evidence from the field was convincing and the Sturmgeschütz III was considered to be an instant success. The Sturmgeschütz III was again deployed and proved its effectiveness in the lightning campaigns which swept the Wehrmacht through the Balkans and into Greece. In June 1941 the Sturmgeschütz III was deployed in greater numbers for the attack on Russia and, to a far more limited extent, they also saw action with Rommel in North Africa.

The Russian front was to become Hitler's nemesis and the first signs of impending doom came in the form of the T-34 and the KV-1. Increasingly the Sturmgeschütze III were being called upon to combat this new breed of Russian armour, but the short barrelled main armament was not adequate to develop the high muzzle velocity required for armour piercing rounds, rendering them unsuitable for this role.

The situation in the field was increasingly desperate, but in late 1942, the problem was to an extent solved when the ineffective

A Stug III struggles to cope with the unexpectedly severe weather during the first winter of the war on the Russian front.

Stug III in the streets of Kharkov; October 1941.

short barrelled 75mm gun was replaced with the long barrelled high velocity StuK 40 L/48. Equipped with this gun, the StuG III soon proved itself to be a highly effective tank destroyer. Its extremely low silhouette meant that the StuG III was capable of lying in wait to ambush Soviet tank formations. As a result of the constant programme of modification and upgrading, which lasted throughout the war, variations of the StuG III, and its later rival the StuG IV, served very effectively as assault guns, mobile artillery and tank destroyers.

The StuG III was conceived to fill an obvious gap for a self-propelled artillery piece which could be brought right up to the front lines in support of fast moving infantry. However, it was not initially clear which arm of the Wehrmacht would actually field the new weapon. The Panzerwaffe was the obvious choice as the natural user of tracked fighting vehicles, but in 1939 Germany's tank arm was undergoing a rapid and ambitious expansion programme which was focussed on developing tanks as a breakthrough weapon. In 1939 there were simply no resources to spare for the formation of StuG units which were intended to perform a secondary battlefield role as viewed from the narrow perspective of the Panzerwaffe.

The infantry branch lacked the knowledge, infrastructure and the resources to satisfactorily implement the full capabilities of the new machines. It was therefore agreed, after much discussion, that the StuG IIIs would best be employed as part of the Heer's artillery arm. The crews of the StuGs were therefore artillerymen rather than tank men and retained the grey uniform as opposed to the Panzer black of the tank men.

From the outset the Sturmgeschütze were therefore organised along artillery lines initially formed into batteries, then into battalions (which were later designated as brigades in order to try and exaggerate the numbers in an attempt to deceive Allied intelligence). A report of the first of the StuG III's in combat was promulgated to American troops by the Military Intelligence Service in December 1941 some six months after the commencement of Operation Barbarossa.

An early production model StuG III in action during the advance into Russia.

THE CONTEMPORARY VIEW NO.I

GERMAN ARMORED ASSAULT ARTILLERY

SOURCE

This bulletin is based upon the report of an American official observer in Berlin. The translated article, which deals with the employment of a battery of armored assault artillery of the "Greater Germany" Infantry Regiment on the French-Luxemburg border, originally appeared in Die Woche, a German weekly magazine.

Intended propagandistic effects should not be overlooked.

CONTENTS

1. Translation

2. Comments of Official Observer

I. TRANSLATION

"A motorized platoon, with 2 antitank guns attached, constituted the leading element of our advance guard as we marched west from Vance, which is 20

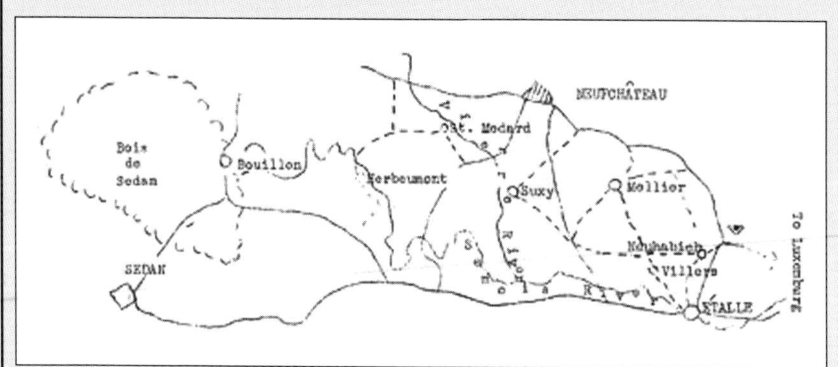

Figure 1. Area around Étalle and Neufchâteau.

Armored tanks and assault guns in the streets of Kharkov.

miles west of Luxemburg, to Étalle (figure 1). As the vehicles approached Étalle on May 10, they encountered hostile armored scout cars, and during the ensuing engagement a report was received at regimental headquarters that Villers was occupied by French cavalry. The 2d Battalion was accordingly ordered to attack Villers immediately. For 3 hours they advanced toward the village, meeting increasing resistance, and were finally stopped at the eastern edge by strong hostile fire.

"Meanwhile the 1st Battalion, with the armored assault artillery battery attached, had arrived at Neuhabich, where the battalion commander ordered a rifle company to make contact with the 2d Battalion. Advancing slowly south from Neuhabich, the rifle company finally reached Villers, where it also met heavy resistance. The company commander, after considering the situation, sent the following oral message to the rear: 'Assault

SdKfz 253 with headlights removed and round plate covering the hole. The SdKfz 253 command vehicle proved inadequate for the task and was phased out from 1942 onwards.

battery to the front!'

"The 3d Platoon of the armored assault artillery battery dashed forward to engage in its first fight. The platoon commander, in his command vehicle, was followed by Assault Guns No. 5 and No. 6. The platoon encountered no resistance until it arrived at the center of town, where it received heavy machine-gun fire. Two rounds from each of the assault guns silenced the machine guns.

"Assault Gun No. 6 went into action, firing at the nearby buildings. One shell exploded in a courtyard among some French cavalry horses. The animals which were uninjured galloped away, frightened by the explosion.

"Assault Gun No. 5 swung into position in the churchyard to silence hostile machine guns which were firing from two windows in a large building close by. The platoon commander ordered the gun commander to fire on this target, and two rounds from the assault gun served to silence them.

"Finally, the enemy evacuated the main street and the center of the town, but machine-gun resistance was renewed at the western edge of the village. Momentarily, it was thought that the assault guns should be sent ahead again. But the riflemen and the partially armored antitank-weapons on self-propelled mounts were able to reduce this resistance unassisted.

"The 2d Battalion remained in Villers during the night. Field kitchens were moved up, the men were fed, and medical personnel cared for the wounded. The 3d Platoon of the assault battery obtained some rest just in rear of the front line, the men sleeping in their vehicles. The next morning, at 5 a.m., the advance guard and the 3d Platoon of the assault battery moved out toward Mellier.

"The armored assault guns soon reached a destroyed bridge across a tributary of the Semois River. The pioneers, although hard at work, had not yet completed their task here; but the guns managed to ford the river. The regimental commander, in order to get up to the front, took a seat in an assault battery munitions vehicle.

"After fording the stream, the assault guns came to a barricade of tree trunks which obstructed the road leading up a slope in one of the southern spurs of the Ardennes Forest. The driver of Assault Gun No. 5, however, stepped on his accelerator, dashed against the obstacle, and opened the way. So far, no enemy had been encountered.

"The infantry was in the lead as the advance guard moved through Mellier into a beechwood forest beyond that town. Resistance was encountered at 10:30 a.m. at a clearing in the woods. The 1st Battalion, upon emerging into the clearing, was fired upon from the direction of Suxy. The leading company deployed promptly and, supported by an antitank platoon, began to advance, finally being checked at the stream just west of the town. The regimental and battalion commanders, accompanied by certain members of their staffs, observed the action from high ground east of Suxy. Intense activity prevailed at the command posts. Heavy weapons were ordered up; tasks were assigned, and positions designated.

"As the heavy infantry weapons and armored assault guns were heard approaching from the rear, the battalion commander, in a quick decision, signalled his advancing reserve company to turn off and attack in the new direction.

"Five minutes after the heavy weapons arrived, they opened fire. In the meantime, the armored assault artillery battery continued to the front to assist the leading rifle companies. The riflemen slowly worked their way ahead, pressing hard against the enemy, driving him off of the high ground to the right front. Finally, one of the assault guns moved up on to this commanding terrain and quickly fired 11 rounds at a range of 800 yards into a battery of enemy horse artillery going into action. The assault gun itself, however, was then taken under fire by a French antitank battery.

"In the meantime, the German artillery opened fire and the battalion began to advance across the Vierre River. As usual, all the bridges had been destroyed and all the trucks had to be left behind, although the water was no obstacle for the infantry and the armored assault artillery.

"After crossing the river, the advance, was checked again by resistance coming principally from a fortified house which stood along the route of advance. Assault Gun No. 5 went into action against this house. The first round hit the lower left window; the second entered the attic window; the third went over the house but exploded among some retreating Frenchmen.

"By 5:30 p.m. all resistance in this vicinity had been overcome. The French reconnaissance battalion, which had attempted to stop the regiment, was completely destroyed.

"The advance continued, but the next 10 miles could be covered only by foot, for the trucks could not be moved across the river. The day's objective, however, was reached at 9 p.m.

"The performance of the armored assault artillery battery, in its initial engagements at Villers and Suxy, completely won the confidence of the infantrymen. In addition to giving support to the foot soldier in battle, this self-propelled artillery was also utilized in carrying light machine guns and mortars and in towing ammunition carts.

"On the next morning, May 12, the regiment moved

through St. Medard and Herbeumont.

"On the following day, May 13, the regiment left Belgian soil, marching through Bouillon into the Bois de Sedan, and on the next morning it forced a crossing over the Meuse at Sedan, thereby clearing the road to the north for the oncoming panzer division."

2. COMMENTS OF OFFICIAL OBSERVER

a. The personnel of the "Greater Germany" Infantry Regiment is especially selected. Initially, the bulk of men of this organization came from the Berlin guard regiment. The regiment is motorized and belongs to an S.S. division.

b. The author indicates that in this particular engagement this assault artillery fulfilled the mission for which it was intended. Conversations with German military personnel and the context of other articles published in German military periodicals confirm the conclusion that this assault artillery gave important and timely assistance to the leading infantry elements on many occasions during the operations on the Western Front in the spring of 1940.

c. Since this weapon is completely armored, it conforms to the commonly accepted definition of a tank. According to published accounts, this weapon, during combat, moved forward from cover to cover, keeping generally abreast of the regimental reserve. When the advance of the leading foot elements was checked by resistance beyond the capabilities of the infantry weapons immediately at hand, the armored assault artillery was ordered forward along with other heavy infantry weapons and sometimes the regimental infantry reserve. When going into action, armored assault artillery vehicles sought suitable covered positions in the front line, from which they delivered direct fire upon observed targets. It is not believed that they ever preceded and cleared the way for the foot elements. Consequently, these weapons, as employed, are not comparable to accompanying tanks.

d. It is probable that if the defending French forces had been liberally equipped with antitank mines and antitank weapons, they could have neutralized the efforts of the German armored assault artillery.

e. The action east of Suxy is in accordance with the German principle that attacking infantry seeks primarily to seize commanding terrain, not only for observation and the employment of infantry weapons, but also for the advantageous use of artillery.

It is noted that in the attack east of Suxy, the commander of the 1st Battalion saved time by merely signalling to the commander of his reserve company to turn off his route of advance and attack in a new direction. Details of the action contemplated for this company could be furnished later by the battalion commander or his representative.

CHAPTER 2

THE DEVELOPMENT OF THE STURMGESCHÜTZ

The Sturmgeschütz originated from the German experience in World War I when it was discovered that during the 1918 offensives on the western front, the fast moving Stoßtruppen infantry lacked the means to effectively engage strongpoints. The horse drawn artillery of the time was highly vulnerable to machine gun and small arms fire and not sufficiently mobile enough to keep up with the advancing infantry, who therefore lacked the means to destroy bunkers, pillboxes, and other minor obstacles with direct-fire artillery fire.

Although the problem was well-known in the post-war German army, it was not formally addressed until General Erich von Manstein tackled the issue in 1935. He gave the Sturmgeschütz idea its momentum and as a result is now considered the father of the

The Sturmgeschütz viewed from the enemy point of view. This propaganda photograph was probably taken in Russia during 1942.

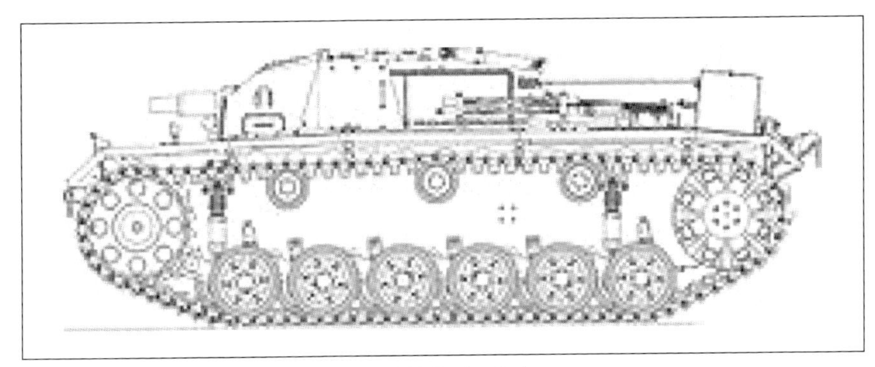

StuG III, Ausf. A

Sturmartillerie. The initial proposal which led to the introduction of the StuG III was drafted by von Manstein and was first submitted to General Ludwig Beck in 1935. Manstein advocated that the fast moving assault tactics of the Blitzkrieg era would demand the support of Sturmartillerie units which could keep up with the infantry and provide direct-fire support role for the infantry in close combat situations.

Manstein's proposal resulted, on June 15, 1936, in Daimler-Benz AG receiving the order to develop an armoured infantry support vehicle capable of mounting a 75 mm artillery piece. The gun was mounted in a fixed, fully integrated casemate superstructure which allowed only a limited traverse of a maximum of 25° but provided fully enclosed protection for the four man crew. The height of the vehicle was not to exceed that of the average man.

Daimler-Benz AG wisely decided to use the chassis and running gear of its recently designed Pz.Kpfw. III medium tank as a basis for the new vehicle. Prototype manufacture was passed over to Alkett, which produced five examples in 1937. This was the experimental 0-series StuG based upon the Pz.Kpfw. III Ausf. B. These prototypes featured a mild steel superstructure and Krupp's short-barrelled 75 mm StuK 37 L/24 cannon. This model was known as the Sturmgeschütz Ausführung A.

The vehicles of the Sturmgeschütz series were cheaper and faster to build than contemporary German tanks; at 82,500 RM, a StuG III Ausf G was cheaper than a Panzer III Ausf. M, which cost 103,163 RM. This was due to the omission of the turret, which greatly simplified manufacture and allowed the chassis to carry a larger gun than would otherwise be the case. By the end of the war, 10,619 StuG IIIs and StuH 42s had been built.

From the British perspective The Stug III made little impact on the campaign in the west. However the Western Allies began to receive information on the Stug III as a result of their own experiences in North Africa and from Soviet Sources. The capture of a Stug III in north Africa led to the first Allied intelligence report on the Stug III was entitled "German 75-mm Assault Gun" and appeared in the US intelligence magazine Tactical and Technical Trends No 7, on 10th September 1942.

This photograph demonstrates the building of Sturmgeschütze and Panzer Mark IIIs side by side in the Alkett works in early 1941.

THE CONTEMPORARY VIEW NO.2

GERMAN 75-MM ASSAULT GUN

This assault gun is a self-propelled gun mounted on a standard Mark III tank chassis. In 1940 a relatively small number took part in the Battle of France and it was first used extensively in the summer of 1941, when it played an important tactical role in the first battles on the Russian front.

The guns are organized into independent battalions, although it is now possible that they are organic within the motorized and Panzer divisions and are attached to front-line infantry divisions. Normally only direct fire is used.

An assault gun captured in the Middle East is described below.

The gun and mount weigh about 20 tons.

The gun itself is the short-barreled 75-mm tank gun originally mounted in the Mark IV tank. The range drum is graduated for HE up to 6,550 yards and for AP up to 1,640 yards. Elevation and traverse are hand-operated. Some other details are in the table overleaf.

It is believed that this low-velocity gun is being replaced by a high-velocity 75-mm gun with a reported length of bore of about 43 calibers. The Germans are also apparently making a similar change in the armament of the Mark IV Tank.

As stated above, the hull is that of the standard German

Length of bore	23.5 cals.
Muzzle velocity (estimated)	1,600 f.s.
Elevation	20°
Depression	5°
Traverse	20°
Weight of projectiles	
HE	12 lb. 9 oz.
Smoke	13 lb. 9 oz.
AP (with ballistic cap)	13 lb. 9 oz.
AP (hollow charge)	not known
Estimated penetration of AP (with ballistic cap)	55 mm. (2.16 in.) at 60° at 400 yds.

Technical specifications 75-mm assault gun

Mark III tank with normal suspension system. The turret has been removed. The length is 17 ft. 9 in., height 6 ft. 5 in., and width 9 ft. 7 in. In general the armor is 51 mm. (2 in.) at the front and 32 mm. (1.25 in.) on the sides and at the rear. An added 53-mm plate is fitted to the rear of the front vertical plate, apparently between the driving and fighting compartments, and is braced to the front plate by two 31-mm. plates, one on each side of the opening for the gun. For detailed arrangement of armor plate see accompanying sketch opposite.

The sides of the hull are reported to be vulnerable to the British 40-mm antitank gun at 1,500 yards, but this gun can penetrate the front only at very short ranges, and even then only the driving compartment.

The engine is a Maybach V-12-type rated at 300 horsepower. The gears provide for six speeds, and steering is hydraulically controlled. The capacity of the gasoline tank is 71 gallons, which is consumed at the rate of about 0.9 miles per gallon at a cruising speed of 22 miles per hour. The radius of action is about 70 miles, the maximum rate of speed about 29 miles per hour.

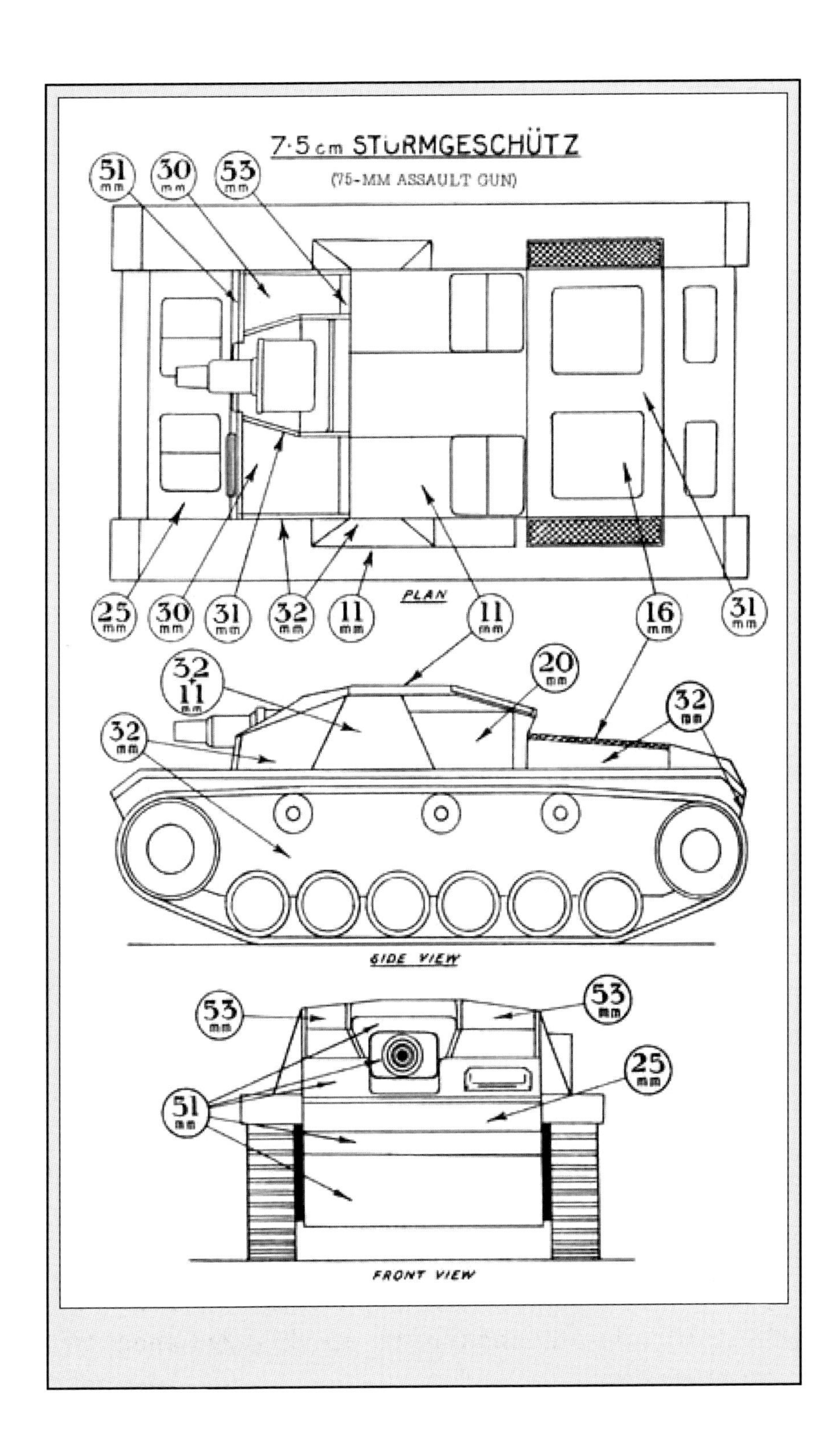
7·5 cm STURMGESCHÜTZ
(75-MM ASSAULT GUN)
51 mm
30 mm
53 mm
25 mm
30 mm
31 mm
32 mm
11 mm
PLAN
11 mm
16 mm
31 mm
32 11 mm
20 mm
32 mm
32 mm
SIDE VIEW
53 mm
53 mm
25 mm
51 mm
FRONT VIEW

As in German tanks, this vehicle is equipped to carry extra gasoline in a rack on the rear of the vehicle, which should hold about 10 standard 5-gallon gasoline cans.

The captured vehicle contained metal boxes for 44 rounds of ammunition, and 40 rounds were stacked on the floor at the loader's station. Ammunition is also carried in an armored half-track which tows an armored ammunition trailer. There was also a rack for 12 stick grenades, and the usual smoke-candle release mechanism for 5 candles was fitted to the rear. For communication there were two radio receivers and one transmitter. For observation a scissors telescope was provided.

As spare parts the 11-mm. sloping plates over the track guard (see sketch) carried two spare bogie wheels on the right side and one on the left side. Two spare torsion rods were also carried, one in each side of the hull above the bogies.

The crew consists of four men - a commander, gunner, loader, and driver.

Assembly workers hard at work producing Sturmgeschtze in January 1942. These machines are equipped with the short-barrelled 75mm which had already proven to be ineffective for many of the tasks faced by the Stug.

As the StuG III was designed to fill an infantry close support combat role, early models were fitted with a low-velocity 75 mm StuK 37 L/24 gun which could be used to destroy soft-skin targets and field fortifications. To the extent that it proved necessary, the main gun was discovered to be capable of dealing with the British and French tanks of the period at reasonably long ranges. However, following the launch of Operation Barbarossa, the Germans encountered the Soviet KV-1 and T-34 tanks and it soon became clear that the relatively low velocity of the main gun was not up to the task of dealing with these stronger armoured vehicles.

Defeating the Soviet tanks required a projectile with much increased muzzle velocity; the speed a projectile achieves at the moment it leaves the muzzle of the gun. Longer barrels give the propellant force more time to develop the speed of the shell. For this reason longer barrels generally provide higher velocities, everything else being equal. Throughout World War II muzzle velocities constantly increased as larger tank guns firing more powerful penetrative ammunition were introduced. Muzzle velocity is also affected by the burn speed, rate of expansion and quantity of the propellant, the mass of the projectile, and the length of the barrel. Faster burning propellant, improved shells and longer barrel length constantly enhanced tank-killing performance on both sides.

As a result of this evolutionary race, the StuG III was equipped with firstly a high-velocity 75 mm StuK 40 L/43 main gun, which required relatively few modifications in order to make the gun work within the framework of the existing design. The up-gunned StuG IIIs made their debut in the spring of 1942 and were deemed a qualified success. The L/43 was something of a stop-gap however, and a short time later the engineers delivered a design for a Sturmgeschütz incorporating the longer barrelled StuK 40 L/48 which began to appear from Autumn 1942 onwards. This was a much more effective anti-tank gun and was used to equip the Sturmgeschütz batteries for the rest of the war. The versions of the Sturmgeschütz equipped with this weapon were known as the Sturmgeschütz 40 Ausführung F, Ausf. F/8 and Ausf. G.

The Stug III was originally known simply as the Sturmgeschütz, but when production difficulties at the Alkett factory halted production of the Stug III in late 1943, the StuG IV, based on Krupp's Mark IV chassis entered production as an interim solution. It became necessary to differentiate between the two vehicles. The Roman numeral "III" was therefore added to the name in order to delineate the early models from the later machines based on the Panzer IV. Posterity now knows the original vehicle as the Sturmgeschütz III or StuG III.

The original StuG III lacked a machine gun for close defence and this deficiency soon became noticeable in Russia. Beginning with the StuG III Ausf. G, a 7.92 mm MG34 could be mounted on a shield on top of the superstructure for added anti-infantry protection from December 1942. Some of the earlier F/8 models were retrofitted with a shield as well. Many of the later StuG III Ausf. G models were equipped with an additional coaxial 7.92 mm MG34. From December 1942, a square machine gun shield for the loader was installed, allowing an MG 34 to be factory installed on a StuG for the first time. F/8 models had machine gun shields retro-fitted from early 1943. The machine gun shield for the loader was later replaced by rotating machine gun mount that could be operated by loader inside

A Sturmgeschütz commander and supporting officers review a tense situation Russia 1943.

the vehicle by a periscope. On April 1944, 27 of them were being field tested on the Eastern front. Favourable reports led to the installation of these "remote" machine gun mounts from summer of 1944.

THE CONTEMPORARY VIEW NO.3

REMOTE-CONTROLLED MACHINE GUN FOR ASSAULT GUNS AND TANK DESTROYERS

Extracted from the US Intelligence Bulletin, May 1945

The Germans are equipping tank destroyers and assault guns with an indirect-laying and indirect-aiming device so that the personnel of these heavy armored vehicles can put up an improved defense against close-in attacks without exposing themselves. The new device consists of a standard light machine gun, the M.G. 34, mounted on a hollow-column base on top of the armored fighting compartment. Controls are fitted so that the gunner can aim, elevate, and traverse the gun from his seat inside the vehicle. The gun's shield is so designed that a standard 50-round drum may be attached to the left of the gun's receiver.

An armored shield protects the mechanism of the gun and also the mount.

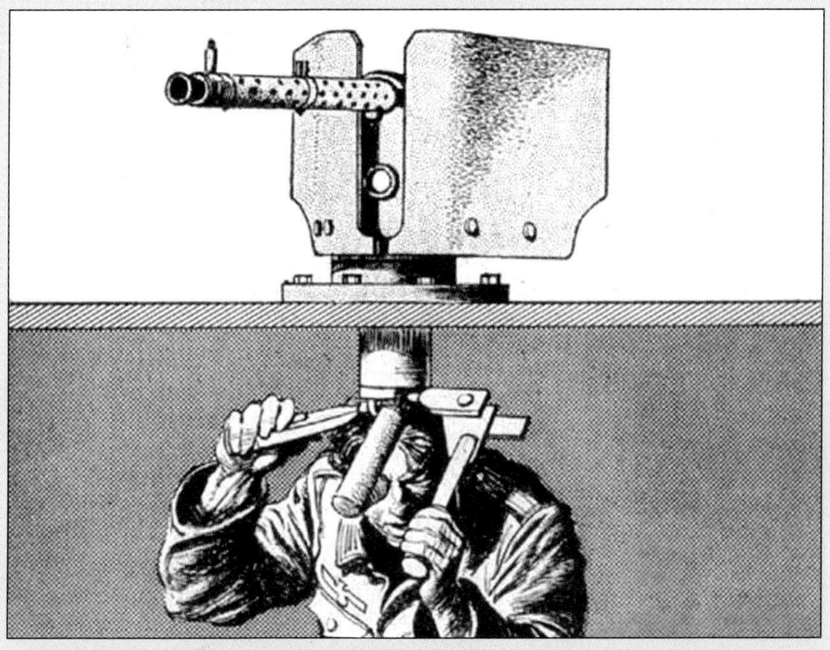

The gunner remains inside the armored fighting compartment. With one hand he controls the traverse; with the other, the elevation. He aims by means of a periscope. To reload, he must open a hatch

This remote-controlled gun differs from previous forms of superstructure-mounted machine guns in that each of the conventional types must be operated by a man standing with his head and shoulders exposed above an open hatch.

Although some protection is afforded by a folding shield which faces forward, the conventional mounts permit forward fire only. The new type of mount is designed to give protection against attack from sides and rear, as well, and to supplement the fire of the bow gun now fitted on the latest German tank destroyers.

An early version of the StuG III moving up in support of the infantry through a blazing Soviet landscape.

A rare study of a pair of Sturmgeschütz and a command vehicle in action in Russia 1941.

INITIAL DEPLOYMENT AND ORGANIZATION

The first 24 StuG III Ausf A's produced in early 1940 and were used to equip the Sturmartillerie Batteries numbered 640, 659, 660 and 665. They first saw service during the French Campaign. Each battery was organised consisted of six assault guns organised into three platoons comprising two assault guns each. Sturmartillerie Battery 640 was subsumed into Infantry Regiment Grossdeutschland and was renamed the 16th Sturmartillerie Battery. The last six assault guns were issued to the SS Sturmartillerie battery of Leibstandarte SS Adolf Hitler division. Two more batteries were formed in 1940 and these were numbered the 666th and 667th batteries, but were formed too late to see action in France.

In August of 1940, Sturmartillerie units were reorganized into Abteilungen or battalions with 18 assault guns each, organised into in three batteries of six assault guns in each battery.

In early 1941 the name Sturmartillerie disappeared and the battalions were officially renamed Sturmgeschütz Abteilungen, and

Re-supply in the open during the advance of Army Group North. The relaxed attitude suggests that the Red air force posed no threat at this stage.

the batteries were redesignated as Sturmgeschütz Batteries. The composition of the units remained unchanged and each Abteilung still consisted of three batteries of six guns each.

In the spring of 1941, all Sturmgeschütz Batteries had the number of assault guns increased from six to seven.

In 1942, with the introduction of long-barrelled StuG III (75mm L/43 and L/48), the existing Sturmgeschütz Abteilungen were expanded and number of assault guns was increased to 28 per battalion. Each battalion still had three batteries but the number of assault guns in platoons was increased to three with specialist command vehicles making up the additional numbers.

In November of 1942, Sturmgeschütz Abteilungen were reorganised yet again and the number of assault guns was increased to 31 per battalion with three additional assault guns for battery commanders. This type of organization was often referred to as Sturmgeschütz Brigade, and remained in use until the end of the war.

In June of 1944, a further re-organization scheme was introduced – once again the term Sturmartillerie was brought back into use and the formations were now called Sturmartillerie Brigades. In theory each brigade now consisted 45 assault guns, comprising 33 StuG III/ IV (75mm L/48) assault guns and 12 Sturmhaubitze 42 (105mm L/28) assault howitzers. The brigade was organised into three batteries with two StuG III's for each battery command, while each battery had two platoons of four StuG III's and one of four StuH 42s. This organization scheme was used alongside the Sturmgeschütz Brigade scheme to the end of the war. In practice, these ideals were hardly ever achieved and then only highly favoured formations received the full complement.

Towards the end of the war, StuGs (40) were often issued to other units as replacements for tank destroyers and even tanks. Since 1944, StuG III (40) were also used as replacements for PzKpfw III, PzKpfw IV and even PzKpfw V Panther in Panzer Abteilungen.

During the course of war, StuG III assault guns were issued to Sturmartillerie Batteries, Sturmgeschütz Abteilungen, Sturmgeschütz Brigades, Sturmartillerie Brigades, Ersatz (Reserve) Abteilungen and Funklenk (Remote Control) Companies. StuG III assault guns served on all fronts of World War II to the end of the war. Only the elite Wehrmacht (e.g. Grossdeutschland) and Waffen SS (e.g. LSSAH, Das Reich, Totenkopf) divisions had Sturmgeschützbrigaden as permanent part of their divisions.

THE CONTEMPORARY VIEW NO.4

75-MM ASSAULT ARTILLERY

Extracted from the US Intelligence Bulletin, July 1943

The German 75-mm assault gun is a weapon comparable to the U. S. 75-mm and 105-mm self-propelled guns. The gun and mount weigh about 20 tons. The maximum speed across country is about 7 miles per hour; on roads, about 22 miles per hour. It can average about 15 miles per hour. On normal roads its radius of action is about 100 miles; across country, about 50 miles. To move an assault-gun battery 100 kilometers (about 65 miles) requires 4,000 liters (about 1,050 gallons) of gasoline. The range of the 75-mm short-barreled tank gun, with which this weapon was originally equipped, is about 6,000 yards.

Apparently there are now three types of German assault guns in service: the short-barreled 75-mm tank gun, with a bore 23.5 calibers in length; the long-barreled 75-mm tank gun, with a bore 43 calibers in length; and an intermediate gun which seems to be a 75-mm gun with a bore 30 calibers in length. It seems probable that the long-barreled 75, which is the principal armament of the new Pz. Kw. 4 tank, may be primarily an antitank weapon, while the intermediate gun will take the place of the old short-barreled 75 as a close-support weapon.

A 1940 German document states that the assault gun "is not to be used for antitank purposes, and will only engage enemy tanks in self-defense or where the antitank guns cannot deal with them." However, a 1942 German document states that "the assault gun may be used successfully against armored vehicles and light and medium tanks." This apparent contradiction can perhaps be explained by the fact that prior to the invasion of Russia in 1941, this weapon had been used in limited numbers. Experience in Russia may have shown that it could be used successfully against tanks, although Russian sources refer to it as an infantry support weapon, essentially. Perhaps a more logical explanation lies in two German technical developments since 1940, namely: hollow-charge ammunition, which is designed to achieve good armor-piercing performance at relatively low muzzle velocities, and the reported replacement of the short-barreled, low-velocity 75-mm with the long-barreled, high-velocity 75-mm gun on some of the newer models.

The following information about German assault artillery is a condensation of a recent article in "Red Star," the official Soviet Army publication, and deals with only one of the three types—the short-barreled 75-mm.

The Germans make extensive use of self-propelled guns as assault artillery. Their most important mission is to destroy the opposition's antitank and heavy infantry weapons. The German self-propelled mount under discussion is a Pz. Kw. 3 chassis armed with a short-barreled 75-mm gun, which has a semiautomatic breech block. The gun's traverse is limited. The armor on the front and sides of the vehicle has thicknesses of 50 mm and 30 mm, respectively. The top and rear of the gun carriage is open. The speed of the self-propelled gun is about 31 miles per hour, and its range is about 84 miles. The gun's initial muzzle velocity is about 1,389 feet per second. The gun carries 56 rounds. The ammunition is fixed and consists of the following types: high-explosive, armor-piercing, and smoke.

The gun crew consists of a gun commander, a gunner, a loader, and a driver. Two self-propelled guns make up a platoon. The platoon commander's vehicle is equipped with signal flags, rocket pistols, a two-way radio, and a speaking tube for communication between the commander and his gunner and driver. The radius of the radio is about 2 1/2 miles when the vehicle is at the halt, and from 1 1/4 to a little less than 2 miles when it is moving. The second vehicle in the platoon has only a receiving set and signal flags.

There are three platoons in a battery, as well as a separate gun for the battery commander, three armored vehicles with supplies, and an ordinary supply truck. In a battalion (the largest unit) there is a headquarters, a headquarters battery, and three firing batteries. The battalion commander has a gun under his own personal command. According to the German table of organization, the battalion of assault guns is an independent unit and is part of the GHQ artillery pool. The assault artillery battalion can be placed under the command of an infantry commander or tank unit commander, but not under an officer of lower rank than regimental commander. It is important to note that if an assault-gun battery has the necessary supplies to permit it to take care of itself, it may assume an independent role, apart from that of the battalion.

Assault batteries, which are assigned a limited number of targets, have the mission of supporting the attacks of the infantry, and of destroying the opposition's heavy infantry weapons and strong points disclosed during the course of the attack. In supporting tank attacks, the self-propelled artillery assumes some of the normal tasks of the heavier tanks, including the destruction of antitank guns.

The assault artillery never serves as antitank artillery in an attack; only in self-defense does it open fire at short range, shooting armor-piercing shells against tanks. Its shell has almost no effect against heavy tanks.

The battery is part of the combat echelon, and marches ahead of the trains. All seven guns and three armored supply vehicles are in this echelon. In deploying for battle the guns come first, moving abreast toward the front and ready for instant action. The guns of the platoon commanders are on the flanks. The battery commander is stationed to the rear, in a position which is dictated by the type of firing and the terrain. Behind him, the supply vehicles move by bounds from one protected position to another.

If a position lacks cover, these vehicles follow at a considerable distance, maintaining radio communication with the rest of the battery.

In carrying out its special task of facilitating an infantry breakthrough into the rear of the opposition's defenses, the assault battery may follow one of two methods of maneuver: the battery may take part in the initial assault, or it may be held in reserve and not committed until the hostile dispositions have been discovered. In all instances the battery cooperates closely with the supported infantry battalion or company.

Assault guns use direct fire. To achieve surprise, they move forward stealthily. In supporting an infantry attack under heavy enemy fire, assault guns halt briefly to fire on target, which offer the greatest danger to the infantry. The assault guns fire a few times, and then disappear to take part in the battle from other positions. When an assault artillery battalion is attached to an infantry division cooperating with Panzer units in an attack, the battalion's primary mission is to destroy the hostile antitank defenses. If the battalion is supporting tanks in a breakthrough, its batteries seek positions permitting good observation. In other cases each battery moves into the attack after the first wave of tanks, and as soon as the latter encounters opposition, the assault guns cover them with protecting fire. It is believed that the Germans regard close cooperation between the assault battery and the first echelon of tanks as essential in effecting a quick destruction of antitank defenses.

If hostile tanks counterattack, the German antitank guns engage them, and the assault artillery unit seeks to destroy the hostile guns which are supporting the attacking tanks. When the German antitank artillery is unable to stop the hostile tanks, as a last resort, the self-propelled assault guns engage the tanks, opening fire on them with armor-piercing shells at a distance of 650 yards or less.

In the pursuit, the assault guns give the infantry close

A study of a column of Sturmgeschütze armed with the long-barrelled 75mm gun moving into action.

support to strengthen the latter's fire power.

The most important role of the assault battery in defense appears to be in support of counterattacks. However, in special instances, they have been used as artillery to reinforce the division artillery. When an assault battery is to support a counterattack, it is freed from all other tasks. The battery, knowing the limits within which the counterattack will operate, acts just as it would in supporting an infantry attack. Assault-battery officers and infantry commanders jointly make a careful reconnaissance of the area in which the counterattack is to take place.

The most vulnerable points of a German self-propelled assault gun, according to the Russians, are the moving parts, the rear half of the fighting compartment, the observation apparatus, and the aiming devices.

The Russians contend that their antitank rifles and all their artillery guns, beginning with their 45-mm cannon, are able to fight successfully against the German assault guns. Heavy losses of self-propelled guns, the Russians say, have greatly weakened the German Army's aggressiveness in the attack and tenacity in the defense.

BEFEHLSWAGEN - COMMAND VEHICLES

As we have seen the Sturmgeschütze were conceived as infantry support vehicles and were originally part of the artillery branch. In the early stages of their deployment they were intended to be directed by a the platoon leader. His vehicle however was not a StuG III, the leader was conveyed in a specialist vehicle known as the leichter, gepanzer Beobachtungskraftwagen (light armored observation half track) Sd.Kfz.253.

The superstructure featured a fully enclosed closed crew compartment and a flat cupola which was off-set slightly to the left of centre and mounted on the roof. The cupola hatches opened to the side and each contained a purpose designed slot to allow scissor periscopes to be deployed from inside the vehicle. Behind the cupola, a rectangular hatch was installed (hinged on the right side). A wooden trough ran along the right top of the superstructure to protect the 2 meter antenna in the down position. A once piece access door was mounted in the rear left side of the superstructure. The 253 carried two Fu. 15 receivers and a Fu. 16 transceiver. Also carried was a backpack radio for dismounted action.

Delivery of the Sd.Kfz.253 began in August, 1940 and the vehicles saw action in the Balkans, Greece, North Africa and Russia. Production ceased in June 1941 in favour of the improved leichte Beobachtungspanzerwagen (light observation half track) Sd.Kfz.250/4. In total 285 Sd.Kfz.253's were delivered.

Sd.Kfz.250/4

The successor to the Sd.Kfz.253's was the Sd.Kfz.253's which was produced in a dazzling array of variants. The Sturmgeschütz units were issued with the Sd.Kfz.250/4 variant which, in its more spacious interior, contained an additional radio and observation equipment for enhanced command and control. The 250/4 carried two Fu. 15 receivers which were supported by a two meter rod antenna as well as the Fu. 16 transceiver which was supported by

The Beobachtungskraftswagen Sd.Kfz.253 was based on the chassis and later incorporated into the Sd.Kfz.250. This unusual viewpoint was photographed in the Balkans and shows the Sd.Kfz. 250 crossing a bridge.

the characteristic two metre star antenna. The Sd.Kfz.250/4 was also equipped with a backpack radio.

Sturmgeschütz III Ausf. E

Eventually the StuG abteilungen were led by commanders installed in purpose built StuG III machines in the form of the Ausf. E. This was a modification of the Ausf. D and purpose designed as a platoon command vehicle. The Ausf. E carried two Fu. 15 receivers in its right pannier and a Fu. 16 transmitter in the left pannier. The Commanders were no longer hampered by the limitations of a lightly armoured half-track and could therefore go anywhere the StuGs could go. This produced much more cohesive leadership.

Radios

Equipment	Description	Frequency range (kilocycles)	Aerial	Range	
				Kilometers (Miles)	
Fu. 15	Ultra short wave receiver "h" (UKw.E.h)	23000-24950	2-Meter Rod		
Fu. 16	10 watt transmitter "h" (10 W.S.h)		2-Meter Rod	4	2
				-2.5	-1.3
	Ultra short wave receiver "h" (UKw.E.h)	23000-24950	2-Meter Rod		

CHAPTER 3

THE STURMGESCHÜTZ IN COMBAT

Overall the Sturmgeschütz series of assault guns proved very successful and served on all fronts as assault guns, mobile artillery and tank destroyers. Although Tigers and Panthers have earned a greater notoriety, the assault guns collectively destroyed more tanks at a far lower cost. The high kill rate stemmed from the low silhouette, StuG IIIs which meant they were easy to camouflage and a produced difficult target. Sturmgeschütz crews were selected from amongst the best candidates and were considered to be the elite of the Heer's artillery units. The results certainly justified the effort and Sturmgeschütz units held a very impressive record of tank kills—some 20,000 enemy tanks are thought to have been claimed by StuG units by the spring of 1944. As of April 10, 1945, there were still 1,053 StuG IIIs and 277 StuH 42s left in service. Approximately

A Sturmgeschütz crossing an improvised bridge in Russia during July 1941.

9,500 StuG IIIs of various types were produced until March 1945 by Alkett and a small number by MIAG, this suggests that by April 1945 8,000 StuGs were either lost in combat or destroyed by their crews through lack of fuel.

As we have seen, the StuG assault guns were extremely cost-effective to build compared to the heavier German tanks. In the anti-tank role the StuG was best used defensively with along clear field of fire. The lack of a traversable turret placed the StuG at a severe disadvantage in the assault role and in highly mobile armour battles but as the German forces were generally on the defensive the StuGs were often suited to their role as hidden menace. As the German military situation deteriorated later in the war, proportionately more and more StuGs were built in comparison to regular tanks. In an effort to replace mounting losses the StuG was issued in preference to tanks.

STURMGESCHÜTZE IN FINNISH SERVICE

In 1943 and 1944, the Finnish Army received a total of 59 StuG III Ausf. Gs from Germany and used them in the war against the Soviet Union. 30 of the vehicles were received in 1943 and 29 in 1944. The 1943 batch destroyed at least 87 enemy tanks for a loss of only 8 StuGs and it is a sign of the fighting quality of the vehicles that not all were destroyed as a result of enemy action and some of these were destroyed by their crews to avoid capture. The 1944 arrived too late to be involved in any further actions as the 1944 peace between Finland and Russia concluded Finland's involvement in the war. After the war, some 49 StuGs remained in service and were the main combat vehicles of the Finnish Army until the early 1960s.

100 StuG III Ausf. G were delivered to Romania in the autumn of 1943. They were officially known as TAs (or TAs T3 to avoid confusion with TAs T4) in the Rumanian army inventory. Losses were much higher than in Finland and by February 1945, only 13 Sturmgeschütze were still in use with the Rumanian 2nd Armoured Regiment.

These StuGs in Finnish service gained the nickname "Sturmi". This photograph was taken during the closing stages of the Continuation War.

StuG III in Romanian service in Russia, 1941.

THE CONTEMPORARY VIEW NO.5

TACTICAL EMPLOYMENT OF GERMAN 75-MM ASSAULT GUN

Extracted from Tactical and Technical Trends, No. 19, February 25, 1943

The German 75-mm assault gun (7.5-cm Sturmgeschütz) is a weapon comparable to the U.S. 75-mm and 105-mm self-propelled guns. The gun and mount weigh about 20 tons. Its maximum speed cross-country is about 7 mph, on roads about 22 mph; it can average about 15 mph. On normal roads its radius of action is about 100 miles, cross-country about 50 miles. To move an assault-gun battery 100 kilometers (about 65 miles) requires 4,000 liters (about 1,050 gallons) of gasoline. The range of the 75-mm short-barrelled tank gun (7.5-cm KwK), with which this weapon was originally equipped, is about 6,000 yards.

It is reported that there are now apparently three types of assault guns in service. These are: the Stu.G. 7.5-cm K, mounting the 7.5-cm KwK (short-barreled tank gun--23.5 calibers*); the Stu.G. lg. 7.5-cm K, mounting the 7.5-cm KwK 40 (long-barreled tank gun--43 calibers); and a third weapon, nomenclature at present unknown, which appears to have a 75-mm gun with a bore 30 calibers in length. It seems probable, therefore, that the 7.5-cm KwK 40, which is the principal armament of the new Pz. Kw. 4 (Mark IV tank), may be primarily an antitank weapon, while the latest intermediate gun will take the place of the old Stu.G. 7.5-cm K as a close-support weapon.

While some technical details of this weapon have been known for some time, relatively little information has been available until recently concerning its tactical employment. Two German documents on the tactical use of this weapon have now been received. One is dated May 1940, the other April 1942. The second document is essentially identical in substance with the first, except that the second contains some additional information. Both documents have been combined into one for the present report, and such apparent contradictions as exist are noted in the translation which follows.

Sturmgeschütze of Army Group Centre pictured in 1942.

INSTRUCTIONS FOR THE EMPLOYMENT OF ASSAULT ARTILLERY

A. BASIC PRINCIPLES AND ROLE

The assault gun (7.5-cm gun on an armored self-propelled mount) is an offensive weapon. It can fire only in the general direction in which the vehicle is pointing Owing to its cross-country performance and its armor, it is able to follow anywhere its own infantry or armored troops.

Support for the infantry in attack is the chief mission of the assault gun by virtue of its armor, maneuverability, and cross-country performance and of the rapidity with which it can open fire. The moral support which the infantry receives through its presence is important.

It does not fire on the move. In close fighting it is vulnerable because its sides are light and it is open-topped. Besides, it has no facilities for defending itself at close quarters. As it is not in a position to carry out independent reconnaissance and fighting tasks, this weapon must always be supported by infantry.

In support of an infantry attack, the assault gun engages the enemy heavy infantry weapons which cannot be quickly or effectively destroyed by other weapons. In support of a tank attack, it takes over part of the role of the Pz. Kw. 4, and deals with enemy antitank guns appearing on the front. It will only infrequently be employed as divisional artillery, if the tactical and ammunition situation permits. Assault artillery is not to be included in the divisional artillery fire plan, but is to be treated only as supplementary, and to be used for special tasks (e.g., roving batteries). Its employment for its principal tasks must always be assured.

B. ORGANIZATION OF THE ASSAULT ARTILLERY BATTALION AND ITS BATTERIES

The assault gun battalion consists of battalion headquarters and three batteries. The battery has six guns--three platoons, each of two guns. The command vehicles for battery and platoon commanders are armored. They make possible, therefore, movement right up to the foremost infantry line to direct the fire.

C. PRINCIPLES FOR EMPLOYMENT

(1) General

Assault gun battalions belong to GHQ artillery. For the conduct of certain engagements, battalions or separate batteries are attached to divisions, or to special task forces. The division commander should attach some or all of the assault artillery batteries under his control to infantry or tank units; only in exceptional circumstances will they be put under the artillery commander. Transfer of batteries from support of one unit to another within the division can be carried out very quickly in the course of a battle. Close liaison with the batteries and within the batteries is of primary importance for the timely fulfillment of their missions. The assault artillery fires from positions in open ground, hidden as far as possible from ground and air observation. Only when employed as part of the divisional artillery will these guns fire from covered positions.

Splitting up of assault-gun units into small parts

An artist's impression of the problems of keeping the Sturmgeschütz re-supplied in action. The limited storage capacity meant that these precious machines had to be constantly re-stocked with ammunition, even under fire.

(platoons or single guns) jeopardizes the fire power and facilitates enemy defense. This should occur only in exceptional cases when the entire battalion cannot be employed, i.e., support of special assault troops or employment over terrain which does not permit observation. If employed singly, mutual fire support and mutual assistance in case of breakdowns and over rough country are not possible.

As complete a picture as possible must be obtained of the enemy's armor-piercing weapons and the positions of his mines; hasty employment without sufficient reconnaissance might well jeopardize the attack. Premature deployment must also be avoided. After an engagement, assault guns must not be given security missions, especially at night. They must be withdrawn for refuelling, overhauling, and resupply. After 4 to 5 days in action, they must be thoroughly serviced. If this is not possible, it must be expected that some will not be fit for action and may fall out. When in rear areas, they must be allotted space near repair shops so that they are readily accessible to maintenance facilities, etc.

Troops co-operating with assault guns must give all support possible in dealing with mines and other obstacles. Artillery and heavy infantry weapons must give support by engaging enemy armor-piercing weapons.

Surprise is essential for the successful employment of assault-gun battalions. It is therefore most important for them to move up and into firing positions under cover, and generally to commence fire without warning. Stationary batteries fire on targets which are for the moment most dangerous to the infantry (especially enemy heavy infantry weapons), destroy them, and then withdraw to cover in order to avoid enemy fire. With the allotment of smoke ammunition (23 percent of the total ammunition issue), it is possible to lay smoke and to blind enemy weapons which, for example, are sited on the flank. Assault artillery renders support to tanks usually after the hostile position has been broken into. In this role, assault-gun batteries supplement Pz. Kw. 4s, and during the fluid stages of the battle direct their fire against enemy antitank weapons to the direct front. They follow very closely the first waves of tanks. Destruction of enemy antitank weapons on the flanks of an attack will frequently be the task of the Pz. Kw. 4.

Against concrete positions, assault guns should be

used to engage casemates with armor-piercing shells. Co-operation with assault engineers using flame-throwers is very effective in these cases.

Assault guns are only to be used in towns and woods in conjunction with particularly strong and close infantry support, unless the visibility and field of fire are so limited as to make use of the guns impossible without endangering friendly troops. Assault guns are not suitable for use in darkness. Their use in snow is also restricted, as they must usually keep to available roads where enemy defense is sure to be met.

(2) Tactical Employment

(a) On the Move

Vehicles on the move should be kept well spaced. Since the average speed of assault guns is about 15 mph, they must be used in leap-frog fashion when operating with an infantry division. Crossing bridges must be the subject of careful handling. Speed must be reduced to less than 5 mph, and the assault guns must keep exactly to the middle of the bridge, with intervals of at least 35 yards. Bridges must be capable of a load of 22 tons. The commander of the assault guns must cooperate with the officer in charge of the bridge.

(1) In the Infantry Division

While on the move, the division commander keeps the assault-gun battalion as long as possible under his own control. According to the situation and the terrain he can, while on the move, place one assault gun battery in each combat team. The attachment of these weapons to the advance guard is exceptional. In general, assault gun batteries are concentrated in the interval between the advance guard and the main body, and are subject to the orders of the column commander. On the march, the battery commander and his party should accompany the column commander.

(2) In the Armored Division

On the move, the assault gun battalion attached to an armored division can be used to best advantage if included in the advance guard.

(b) In the Attack with an Infantry Division

The division commander normally attaches assault-gun batteries to the infantry regiments. On receipt of orders placing him under command of an infantry regiment, the battery commander must report in person to the commander of that infantry regiment. Exhaustive discussion between these two (as to enemy situation, preparation of the regiment for the attack, proposed conduct of the attack, main point of the attack, co-operation with divisional artillery, etc.) will provide the basis for the ultimate employment of the assault-gun battery.

It is an error to allot to the battery tasks and targets which can be undertaken by the heavy infantry weapons or the divisional artillery. The battery should rather be employed to engage such nests of resistance as are not known before the beginning of the attack, and which, at the beginning or in the course of the battle, cannot be quickly enough engaged by heavy infantry weapons and artillery. It is the special role of the assault-gun battery to assist the infantry in fighting its way through deep enemy defense zones. Therefore, it must not be committed until the divisional artillery and the heavy infantry weapons can no longer render adequate support.

The attached battery can be employed as follows:

(1) Before the attack begins, it is located so as to be capable of promptly supporting the regiment's main effort; (or)

(2) The battery is held in the rear, and is only committed if, after the attack begins, a clear picture is obtained of the enemy's dispositions.

Under both circumstances the attachment of the battery, and occasionally of individual platoons, to a battalion may be advantageous.

The commander under whose command the battery is placed gives the battery commander his orders. The latter makes clear to his platoon commanders the specific battle tasks, and shows them, as far as possible on the ground, the targets to be engaged. When in action the battery commander, together with his platoon commanders, must at all times be familiar with the hostile situation, and must reconnoiter the ground over which he is to move and attack. The battery will be so disposed by the platoon commanders in the sectors in which it is expected later to operate that, as it approaches the enemy, the battery, under cover, can follow the infantry from sector to sector. How distant an objective can be given, and yet permit the control of fire by the battery and platoon commanders, is dependent on the country, enemy strength, and enemy action. In close country, and when the enemy weapons are well camouflaged, targets cannot be given to the platoons by the battery commander. In these circumstances, fire control falls to the platoon commanders. The platoons must then co-operate constantly with the most advanced infantry platoons; they remain close to the infantry and engage the nearest targets. The question of dividing a platoon arises only if individual guns are allotted to infantry companies or platoons to carry out specific tasks: e.g., for action deep into the enemy's battle position.

In an attack by tanks attached to an infantry division, the assault-artillery battalion engages chiefly enemy antitank weapons. In this case too, the assault-gun battalion is attached to infantry elements. Well before the beginning of the tank attack, the batteries are disposed in positions of observation from which they can readily engage enemy antitank weapons. They follow up the tanks by platoons, and under special conditions--e.g., in unreconnoitered country-- by guns, as soon as possible. In a deep attack, co-operation with tanks leading an infantry attack is possible when the hostile islands of resistance have been disposed of.

In the enemy tank counterattack, our own antitank guns first engage the hostile tanks. The assault-gun battalion engages the enemy heavy weapons which are supporting the enemy tank counterattack. Only when the antitank guns prove insufficient, do assault guns engage enemy tanks. In this case the assault guns advance within effective range of the enemy tanks, halt, and destroy them with antitank shells.

(c) In the Attack with an Armored Division

In such an attack, the following tasks can be carried out by the assault gun battalion:

Sturmgeschütze roll through the streets of Rome.

(1) Support of the tank attack by neutralizing enemy antitank weapons; (and/or)

(2) Support of the attack by motorized infantry elements.

According to the situation and the plan of attack, the battalion, complete or in part, is attached to the armored brigade, sometimes with parts attached also to the motorized infantry brigade. Within the armored brigade, further allotment to tank regiments is normally necessary. As a rule, complete batteries are attached.

To support the initial phase of the tank attack, assault-gun batteries can be placed in positions of observation if suitable ground is already in our possession. Otherwise the batteries follow in the attack close behind the first waves of tanks, and as soon as the enemy is engaged, support the tanks by attacking enemy antitank weapons.

As the tank attack progresses, it is most important to put enemy defensive weapons out of action as soon as possible. Close support of the leading tanks is the main essential to the carrying out of these tasks.

The support of the motorized infantry attack is carried out according to the principles for the support of the foot infantry attack.

(d) In the Attack as Divisional Artillery

In the attack of a division, the employment of the assault gun battalion as part of the divisional artillery is exceptional. In this role, the assault-gun batteries must be kept free for their more usual mission at all times, and must enter battle with a full issue of ammunition.

(e) In the Pursuit

In the pursuit, assault-gun batteries should be close to their own infantry in order to break at once any enemy resistance. Very close support of the leading infantry units increases their forward momentum. Temporary allotment of individual platoons--under exceptional circumstances, of individual guns--is possible.

(f) In the Defense

In the defense, the primary task of assault artillery is the support of counterthrusts and counterattacks. The assembly area must be sufficiently far from the friendly battle position to enable the assault-gun units to move speedily to that sector which is threatened with a breakthrough. Allotment and employment are carried out according to the plan of the infantry attack. The point of commitment should be arranged as early as possible with the commanders of the infantry units allocated to the counterthrust or counterattack. In the defense as in the attack, the assault-artillery battalion will only be employed in an antitank role if it must defend itself against a tank attack. (Only 12 percent of the ammunition issue is armor-piercing.) If employed as part of the divisional artillery (which is rare), the battalion will be placed under the division artillery commander.

(g) In the Withdrawal

For the support of infantry in withdrawal, batteries, and even individual platoons or guns, are allotted to infantry units. By virtue of their armor, assault guns are able to engage enemy targets even when the infantry has already withdrawn. To assist disengagement from the enemy, tank attacks carried out with limited objectives can be supported by assault guns. Allotment of assault-gun batteries or platoons to rear parties or rear guards is effective.

D. SUPPLIES

As GHQ troops, the battalion takes with it its complete initial issue of ammunition, fuel, and rations. When it is attached to a division, its further supply is handled by the division. The battalion commander is responsible for the correct supply of the battalion and the individual batteries, especially in the pursuit. Every battery, platoon, and gun commander must constantly have in mind the supply situation of his unit. It is his duty to report his needs in sufficient time and with foresight, and to take the necessary action to replenish depleted supplies of ammunition, fuel, and rations.

THE CONTEMPORARY VIEW NO.6

RUSSIAN TANK TACTICS AGAINST GERMAN TANKS

Extracted from Tactical and Technical Trends, No. 16, Jan. 14, 1943

The following report is a literal translation of a portion of a Russian publication concerning the most effective methods of fire against German tanks.

For the successful conduct of fire against enemy tanks, we should proceed as follows:

A. MANNER OF CONDUCTING FIRE FOR THE DESTRUCTION OF ENEMY TANKS

(1) While conducting fire against enemy tanks, and while maneuvering on the battlefield, our tanks should seek cover in partially defiladed positions.

(2) In order to decrease the angle of impact of enemy shells, thereby decreasing their power of penetration, we should try to place our tanks at an angle to the enemy.

(3) In conducting fire against German tanks, we should carefully observe the results of hits, and continue to fire until we see definite signs of a hit (burning tanks, crew leaving the tank, shattering of the tank or the turret). Watch constantly enemy tanks which do not show these signs, even though they show no signs of life. While firing at the active tanks of the enemy, one should be in full readiness to renew the battle against

those apparently knocked out.

B. BASIC TYPES OF GERMAN TANKS AND THEIR MOST VULNERABLE PARTS

The types of tanks most extensively used in the German Army are the following: the 11-ton Czech tank, the Mark III, and the Mark IV. The German self-propelled assault gun (Sturmgeschütz) has also been extensively used.

In addition to the above-mentioned types of tanks, the German Army uses tanks of all the occupied countries; in their general tactical and technical characteristics, their armament and armor, these tanks are inferior.

(1) Against the 11-ton Czech tank, fire as follows:

(a) From the front--against the turret and gun-shield, and below the turret gear case;

(b) From the side--at the third and fourth bogies, against the driving sprocket, and at the gear case under the turret;

(c) From behind--against the circular opening and against the exhaust vent.

Remarks: In frontal fire, with armor-piercing shells, the armor of the turret may be destroyed more quickly than the front part of the hull. In firing at the side and rear, the plates of the hull are penetrated more readily than the plates of the turret.

(2) Against Mark III tanks, fire as follows:

(a) From the front--at the gun mantlet and at the driver's port, and the machine-gun mounting;

(b) From the side--against the armor protecting the engine, and against the turret ports;

(c) From behind--directly beneath the turret, and at the exhaust vent.

Remarks: In firing from the front against the Mark III tank, the turret is more vulnerable than the front of the hull and the turret gear box. In firing from behind, the turret is also more vulnerable than the rear of the hull.

(3) Against the self-propelled assault gun, fire as follows:

(a) From the front--against the front of the hull, the drivers port, and below the tube of the gun;

(b) From the side--against the armor protecting the engine, and the turret.

(c) From behind--against the exhaust vent and directly beneath the turret.

(4) Against the Mark IV, fire as follows:

(a) From the front--against the turret, under the tube of the gun, against the driver's port, and the machine-gun mounting;

(b) From the side--at the center of the hull at the engine compartment, and against the turret port.

(c) From behind--against the turret, and against the exhaust vent.

Remarks: It should be noted that in firing against the front of this tank, the armor of the turret is more vulnerable than the front plate of the turret gear box, and of the hull. In firing at the sides of the tank, the armor plate of the engine compartment and of the turret, is more vulnerable than the armor of the turret gear box.

One very common field practice was to add a layer of concrete over the armour plate above the driver's position. This was intended to improve the protection afforded by the armour plate although exactly how effective this measure proved in the heat of battle will always remain a mystery.

From mid 1943 onwards StuG IIIs were also equipped with schurzen (armour skirts) for further protection against hollow charge weapons which were intended to discharge on passing through the schurzen and be defeated by the main armour of the tank.

During the course of the fighting many early versions of the StuG III were actually recalled to the factory to be re-armed with higher velocity main guns and up-armoured by the simple expedient of bolting on additional armour plates. An unknown number of vehicles were also up-armoured by being strengthened with additional armour plates at field workshops.

It was standard practice that older variants returned for factory repairs were re-equipped with crucial upgrades designed for newer variants creating completely non-standard variants. Models produced between September of 1943 and September of 1944, were also provided with a standard factory applied coating of Zimmerit anti-magnetic paste either in the "waffle plate" or standard pattern. In the same time frame, many StuG IIIs already in service were also coated with Zimmerit in the field.

STURMGESCHÜTZ ACES

During the initial battles the StuG III often came into contact with Soviet Armour of the BT-2 type. The results were encouraging, but also highly misleading. By the middle of July 1941 it was common for StuG platoon commanders to take the field in assault guns rather than command half-tracks, Oberfeldwebel Rudolf Jaenicke commanding StuG number 25, provides a remarkable example of just what could be achieved. As commander of a StuG III platoon he managed to single handedly destroy 12 Soviet BT-2 tanks along with a large number of tractors and other equipment which were loaded on rail carriages.

One of the men decorated for bravery in the Lutchessa Valley fighting was Oberleutnant Peter Frantz, who became the second Wehrmacht soldier to be awarded the Knight's Cross in Russia. During the defensive battles around Tula in December, the 16th Company equipped with StuG III assault guns under his command destroyed 15 enemy tanks in one day and 46 overall. Frantz received

Sturmgeschütz in action during the bitter street fighting for Kharkov.

A StuG III on the outskirts of Stalingrad; September 1942.

his award the following June, and was further rewarded with the Oakleaves in 1943, still serving as an assault artillery commander, holding the rank of Hauptmann.

In the Army Group North Sector probably the most famous Sturmgeschütz ace was the Knights Cross holder Wachtmeister Kurt Kirchner from Stug.Abt.667. It was he who destroyed a total of 30 Soviet tanks during a few days of intense fighting in February of 1942. Kirchner proved that even with the short barrelled main gun it was possible by deft manaouvering and clever use of terrain for the StuG III to overpower the most fearsome adversaries in the form of the KV-1 and the T-34 were generally beyond the capabaility of the Stug III when commanded by average crews, but the advent of the StuG III Aus F with its long barrelled high velocity gun gave the more skilled StuG commanders the opportunity to level the playing field.

One of the most successful early engagements involving the StuG III Ausf F took place in the ruins on the outskirts of the city of Stalingrad in early September of 1942 when a StuG III Ausf F commanded by Oberwachtmeister Kurt Pfreundtner from Stug.

The celebrated and highly decorated Alfred Günther; Sturmgeschütze commander and Knight's cross winner.

Abt.244 lying in ambush destroyed nine Soviet T-34 tanks in just 20 minutes. On September 18th of 1942, Oberwachtmeister Kurt Pfreundtner was presented with the Knights Cross for this remarkable achievement which signalled the StuG III's arrival as a proven tank killer.

A measure of just how effective the long barrelled high velocity gun could be comes from the further example of Hauptmann Peter Franz, also a Knights Cross holder, and the commander of Stug. Abt. "Grossdeutschland". He single handedly destroyed 43 Soviet T-34/76 tanks during the Battle for Borissovka on March 14th of 1943.

Another exceptionally talented commander was Unteroffizier Horst Naumann. In the space of just three days between 1st and 4th of January of 1943, Unteroffizier Horst Naumann from Stug.Abt.184 destroyed 12 Soviet tanks during heavy fighting in the Demyansk area. On January 4th, Naumann was awarded the Knight's Cross for the destruction of a total of 27 Soviet medium and heavy tanks, a remarkable achievement for a Sturmgeschütze commander labouring under the disadvantages posed by the fact that he controlled only a turret-less vehicle.

Perhaps the most formidable of all StuG aces were Franz von Malachowski, and Knight's Cross holder Oberwachtmeister Hugo Primozic of Stug.Abt.667.

Born in 1914, Primozic joined the Reichswehr and fought in the French campaign as a field artillery gunner. In 1942 he joined the 667th Sturmgeschütz battalion which was sent to the Eastern Front. Primozig was most successful ace of the Sturmgeschütz assault gun. He was soon promoted to Oberwachtmeister and Hugo Primozic was Commander of 2/zug. 2/batterie Sturmgeschütz Abteilung 667.

The successful combat career of Hugo Primozic started in earnest when he was issued with one of the first Sturmgeschütz assault guns to mount the long barrelled Sturmkanone 40 Pak L/48 gun. Primozic was undoubtedly one of the most highly skilled, and highly decorated, assault gun commanders. On 15th September 1942 in

Rzhev Russia, his intervention decided the outcome of a fast moving armour battle. Emboldened by his newly enhanced combat efficiency Primozic ordered his platoon to engage a Russian tank brigade and soon knocked-out a series of T-34 tanks. His two-gun troop, fighting in isolation, routed a determined Russian breakthrough attempt at Rzhev, which led to the destruction of 24 Soviet tanks. Under his firm and ice cool control Primozic's detachment of just two guns allowed the Russian tanks to advance to the point where their guns could inflict serious damage. The StuG crews aimed at the most dangerous opponents first, disabling them with rounds fired at suicidally short range. A lucky shot jammed the turret of KV II and Primosic's StuG survived two near catastrophic hits. Despite the huge disparity of numbers Hugo Primozic and his men somehow emerged victorious and ended the day with 24 tank kills. Primosic's own brief account was published in Signal In September 1942 it describes how a large Russian force (including the Stalin Tank Brigade) tried to attack the town and area of Rzhev. "On 15th september we had only two StuGs ready, when the Russians wanted to break through. There was heavy artillery fire, and we had to hide in trenches until the storm was over. When the artillery barrage ended, the first enemy tanks already passed our positions, while we still had to climb aboard our guns."

Moving from position to position, the StuG platoon was fighting both tanks and infantry. Primozic destroyed 24 tanks on that particular day, first those who had broken through, often with one shot. For this action Primozic was awarded the Knight's Cross and promoted to senior sergeant.

Between September to December 1942 Primozic destroyed 60 enemy tanks. After an equally gallant action on the 28th Decemebr 1942 he was awarded the Oakleaves to the Kights Cross, being simultaneously commissioned with the rank of Lieutenant. The propaganda magazine Signal carried a special report on his actions.

"Wachtmeister Primozic destroyed tanks day after day during an enemy offensive. One the fourth day, he had to cover the flank of his division, but during that day he fired his last shell... He had to

Knight's Cross holder Oberwachtmeister Hugo Primozic of Stug.Abt.667.

retreat to escape from encirclement by breaking through... There was another StuG, immobilized, and while the Russians were closing in, Primozic came out of his vehicle to connect both vehicles, and drove both back to their own lines..."

Sturmgeschütz Abteilung 667 achieved a remarkable run of success in its four months in action in the northern and central sectors of the Eastern Front. With just 21 assault guns, it destroyed 468 enemy tanks in that period with Hugo Primozic emerging as the top ace and the Aus F tested and proven in the heat of battle.

An artist's impression from the pages of Signal magazine illustrating the difficulties of moving vehicles under combat conditions. In this evocative image, a Sturmegeschüz of the late war variety is being used to pull another machine out of trouble under fire.

Outside of the Heer, the most notable Waffen SS Stug ace was SS-Sturmbannfuehrer Walter Kniep, who commanded the 2nd Sturmgeschuetz Abteilung of 2nd SS Panzer Division "Das Reich". From July 5th of 1943 to January 17th of 1944, his unit claimed destruction of some 129 Soviet tanks, while losing just two Stugs. Kniep was then awarded the Knight's Cross.

THE CONTEMPORARY VIEW NO.7

ENEMY SELF-PROPELLED GUNS – A SUMMARY OF KNOWN EQUIPMENT

Extracted from Tactical and Technical Trends, No. 25, May 20, 1943

GERMAN SELF-PROPELLED ASSAULT GUNS

(1) 75-mm Assault Gun

A low silhouette, a well-armored body, and a short gun firing forward characterize this assault gun. (See Tactical and Technical Trends, No. 7, p. 9.) The mount is the chassis of the PzKw 3. The suspension consists of six small bogies or each side with three return rollers, a front sprocket, and a rear idler. The vehicle with its weapon is heavy--nearly 20 tons. It is 17 ft. 9 in. long and 9 ft. 7 in. wide, but only 6 ft. 5 in. high. A radius of 102 miles by road and 59 cross country is attained with a 300-hp gasoline motor. The crew is four. Probably, this model is no longer in production.

With its casemate mount, the short-barreled 75-mm gun has a traverse of only 20 degrees, and an elevation varying from minus 5 to plus 20. For HE shell, the gun is sighted to 6,550 yards; for AP, only 1,640. At 500 yards, the penetration is 1.81 inches in 30-degree sloping armor, and 2.16 inches in vertical; at 1,200, it drops to 1.57 and 1.89 inches. The HE shell weighs 12.6 pounds; the AP shell, with cap and ballistic cap, 14.81.

Fig. 6

There is an AP hollow charge of unstated weight, as well as a 13.56-pound smoke shell. In the bins of the carrier, 44 rounds are carried, and about 40 more may be stacked on the floor. A dozen stick grenades (potato mashers) may also be carried clipped on a rack.

(2) 75-mm Medium-Length Assault Gun

This machine is essentially similar to the foregoing, except that a gun 30 calibers long mounted in a large box-like casing has replaced the stubby piece in the earlier model.

(3) 75-mm Long Assault Gun

The third assault gun model is a long-barreled "75" with a prominent muzzle brake. It, too, is mounted on the PzKw 3 chassis (see figure 6). The velocity has been increased to 2,400 f/s, with a resulting increased penetration at 500 yards of 3.5 inches of sloping armor and 4.25 inches of vertical; at 2,000, the penetration is still formidable--244 and 3.03 inches. It is thought that this gun is primarily a tank-destroyer weapon.

THE CONTEMPORARY VIEW NO.8

ASSAULT GUN TACTICS

Extracted from the US Intelligence Bulletin, March 1945

To teach German infantrymen some of the tactics used by assault guns, the Fifteenth German Army outlined the advantages and disadvantages of these self-propelled weapons so that the infantry could have a better understanding of how to cooperate with them in the field.

In reply to the question "What must the infantry know about the assault guns?" the Germans offer these comments:

The assault guns are the strongest weapons against hostile tanks. They engage all your most dangerous enemies, and destroy them or force them to take cover. Assault guns are strong when concentrated, but have no effect when used in small numbers. They are capable of forward fire only, since they have no turrets; therefore they are sensitive to attack from the flanks. This is why the guns must never be employed by themselves, but always in conjunction with infantry. These weapons may be considerably restricted by marshy land, thick woods, and natural or artificial obstacles; moreover, they constitute large targets. They can see and hear little. Even during a battle, the assault guns occasionally must withdraw to cover, and obtain fresh supplies of ammunition and fuel.

A full squad of grenadiers is packed aboard this 1943 Sturmgeschütz.

This brings us to the question of how the infantry should assist the assault guns. Infantrymen must draw the guns' attention to hostile tanks and other targets by means of the signal pistol, prearranged light and flag signals, and shouting.

The infantry must neutralize hostile anti-tank guns.

The flanks of assault guns must be covered and protected by the infantry against hostile tank-hunting detachments, which are always ready to operate against our assault guns. Such protection is especially necessary in built-up areas and in terrain where visibility is poor.

The infantry must warn the assault guns of the proximity of anti-tank obstacles and mines, and must be prepared to guide the guns through such obstacles.

The infantry must take advantage of the guns' fire power to advance in strength via prearranged lanes not under fire.

The assault guns must be given sufficient time for reconnaissance. The guns and the infantry will formulate plans through personal consultation, and will ensure means of communications during battle.

Infantry should not stay too close to the guns, and should not bunch. Instead, deployment is advised, to lessen the danger of drawing hostile fire and to avoid injury by ricochets.

Since the driver of an assault gun has limited vision, infantrymen must keep in mind the danger of being run down, and must move accordingly.

Assault guns are "sitting targets" when they have to wait for the infantry; infantrymen can find cover almost anywhere, but the assault guns cannot.

Since the guns fire at the halt, the infantry must gain ground while the guns are firing.

Although the assault guns are of great assistance when ground is being gained, it is the infantry that must hold the ground.

Since the assault guns must keep their ammunition available for unexpected or especially dangerous targets, the infantry must engage all the targets that it can possibly take on with its heavy and light weapons.

Although the assault guns must withdraw after every engagement, to prepare for the next engagement where their assistance will be required, the infantry will not withdraw.

THE STURMGESCHÜTZE IN RUSSIA

The Sturmgeschütz made a valuable contribution to the success of Operation Barbarossa, but, despite the initial successes, by late 1941 it was apparent that the German armoured force as a whole was seriously under-gunned. The armoured forces of the Wehrmacht were in danger of being swamped by the new Russian T-34 tanks.

We have seen that the short 75mm gun of the Sturmgeschütz was really designed to fire low velocity, high explosive shells in support of infantry formations, and although it used the relatively small chassis of the Panzer III, the fact that it had no turret allowed the Stug to be up-gunned to incorporate the deadly long-barrelled 75mm gun, which could not be fitted to a Panzer III.

With this gun, the Sturmgeschütz was now a match for the T-34, and much more than an infantry support weapon. It was now apparent that, with its low silhouette, the Stug was a much harder target to hit than the Russian T-34. The sloped armour also helped to deflect shots away from the vehicle, and the scales gradually began to tip back in Germany's favour during the fierce tank battles of 1942.

This much-needed upgrade was first incorporated into the Sturmgeschütz Model F in 1942. The long barrelled L/43 gun gave the armour piercing shells fired by the Geschütz a much higher muzzle velocity and therefore a far greater tank killing capability than the 50mm gun of the Panzer III. It was now obvious that the Panzer III had evolved as far as it could and the model was phased out. From 1943 the Panzer III chassis was used exclusively for the manufacture of Sturmgeschütze, production of which continued right up to the last days of the war.

The pressure of battle on the Eastern Front ensured that, almost by accident, the German forces had evolved a very successful example of a new breed of fighting vehicle - the Panzerjäger or Tank Hunters. It was in this role that most Sturmgeschütze were to be employed for the rest of the war, and 20,000 enemy tank kills were claimed by assault gun crews up to the early months of 1944.

Sturmgeschütz and riders during the advance into Stalingrad.

The Sturmgeschütze manual stressed time and time again the need for the gun to be stationary when firing. In this way the highest level of accuracy was achieved. When the commanders heeded this request, the results were devastating.

In a mobile battle, where every second counts, the lack of a turret was a very real disadvantage, but Sturmgeschütze crews learned to adopt defensive tactics designed to lure Russian tanks into carefully constructed killing grounds.

The new tactics certainly worked, and in early 1943 there was another increase in gun power with introduction of the long 48 calibre gun, which gave extra velocity.

It was this new gun which equipped the definitive Sturmgeschütze, the Model G, of which 7720 were eventually manufactured. The sheer number produced is a reflection of just how effective and popular it was in battle.

By comparison to the 11,500 Sturmgeschütze manufactured, only 4,500 Panzer Ills, 6,800 Panzer Ivs and 6,000 Panthers were manufactured in the same period.

While the turreted tanks took on the role of the battlefield rovers,

A Sturmgeschütz of Army Group South in action during December 1943.

designed to forge ahead of the infantry in wide ranging strategic advances, it was the Sturmgeschütze which accompanied the Grenadiers during the dogged fighting on the ground. As the war dragged on, they became the infantry's rock in defence and his armoured fist in attack.

By 1943 the Sturmgeschütz was an indispensable part of both the Panzer Division and the ordinary infantry division. The infantry soon came to know that as long as the Gerschütze were in line, things were in control.

Extraordinary results were achieved by skilled crews in Russia, who sometimes accounted for dozens of Russian machines in a single action. In those actions, a major issue for the Sturmgeschütze commanders was the limited fuel and ammunition capacity of their cramped vehicles which produced a constant need to leave the battlefield. As the war progressed, the soldiers in the front line increasingly took strength from the presence of the Geschütze, so the tactical manual for their employment went to great lengths to stress the importance of the Sturmgeschütz commander keeping his infantry commanders informed that his guns were leaving the line only to re-arm and re-fuel.

It was a standard rule of Sturmgeschütz tactical doctrine that, if possible, not all machines would be withdrawn from the line at the same time, but that they should leave the field in relays, otherwise there was a real danger that the morale of the infantry might collapse if they saw their beloved Geschütze withdrawing from the field.

Despite its undoubted success as a tank killer, there was still an infantry support role for the Sturmgeschütz. The low trajectory 75mm gun was an excellent anti-tank gun, but to reach infantry hiding behind obstacles or other terrain features, a high trajectory Howitzer was still required. A further 1,100 Sturmgeschütze were therefore manufactured with the 10.5cm Howitzer which packed a deadly, high explosive punch which could be used in support of the infantry, either in attack or (more usually) desperate defence. The theory was that for every two troops of 75mm armoured Geschütze to deal with enemy tanks, there would be one troop of howitzer-armoured Geschütze to deal with the infantry. Problems with supplies of vehicles meant that this situation was very rarely achieved in practice.

The tactics set down for cooperation between the two types of assault guns was that the guns armed with 7.5cm guns would target any enemy armour, while the howitzer-armoured vehicles would concentrate on the infantry who accompanied the tanks. In this way, countless Russian attacks came to on the bulwark provided by the Sturmgeschütze battalions.

By 1943 the obvious success of the Sturmgeschütz in the field led the allies to target the Alkett factory responsible tor the production of Stug IIIs for priority bombing. The resultant saturation bombing severely damaged the production factories.

During the period of rebuilding production was switched from Alkett to the Krupp's tank works, but Krupp made Panzer IVs not Panzer IIIs.

During 1943 the 1500 Sturmgeschütze manufactured by Krupp used the Panzer IV tank chassis, combined with the highly successful L/48 gun. These machines, known as Sturmgeschütze IVs, were no less successful than the old Stug IIIs.

By the end of the war over thirty thousand enemy tanks had been destroyed by the Geschütze crews, a ratio of approximately three enemy vehicles to every Stug deployed. It was a mark of the achievement of the guns and the crews that Russian orders forbade their tank commanders from entering into anti-tank duels with the Geschütze head to head, ordering them instead to manoeuvre to find the weaker side and rear armour.

One famous variant of the Stug IV was a real heavyweight, the forerunner of the Sturm Tiger. This was the mighty Brumbar (the grizzly bear, or grumbler), which carried a powerful 15cm Howitzer. The Brumbar had sloping frontal armour 100mm thick and was designed to rumble up to infantry fortifications before firing its massive shell at point blank range.

The Brumbar was originally developed to cope with the close-quarter street fighting at Stalingrad but delays in production meant that they did not see active service until the battle of Kursk in 1943.

After Kursk, Germany was largely fighting a defensive war, but on the few occasions when the Brumbar saw action they acquitted themselves well, and over three hundred were produced by the end of the war.

The eventual production of Sturmgeschütze totalled some 11,500 vehicles, more than any other mark of German fighting vehicle. There were sound reasons for this, as not only was the Sturmgeschütz successful on the battlefield, it was also far less expensive, quicker and easier to manufacture than fully turreted tanks built on the same chassis - a vital consideration for Germany's hard pressed manufacturing industry. German armies on all fronts were desperate for armoured fighting vehicles to stem the flood of Russian and allied armour, so production resources were increasingly switched to the production of Sturmgeschütze, which were well suited to fighting a defensive war.

CHAPTER 4

THE STUG DESIGN PATH

The StuG III Ausf. was an experimental prototype which first appeared in 1937; five were produced. By December 1937 two of these vehicles were in service with Panzer Regiment 1 in Erfurt. The Ausf. 0 series was of unusual construction, it had eight road wheels per side which carried very narrow 360mm tracks. These pre-production vehicles were constructed from just 14.5mm of soft steel superstructure and featured the 7.5 cm StuK 37 L/24 gun. These machines never saw combat for which they were unsuitable as a result of the soft-steel superstructure. However, both machines were still being used for training purposes as late as 1941.

The StuG III Ausf. A: (Sd.Kfz. 142)

This machine went into series production from January 1940-May 1940. Fifty machines were produced and they were first deployed in the Battle of France. The StuG III Ausf. A used the chassis of the Panzer III Ausf. F. Frontal armour was increased to 50mm which was considered to be sufficient to defeat most anti-tank weapons of the time. Daimler-Benz produced the first thirty Sturmgeschütze. An additional twenty were built by Alkett along with a machine which is similar to the Ausf.B. These twenty are often added to the production numbers of Ausf.B. However these twenty vehicles are unique in that they were hybrids produced with Ausf.B, but mechanically more similar to Ausf.A. They were converted from the Panzer III chassis with 30mm armor. 20mm of additional armour was added to meet the spec of 50mm. The 20 vehicles produced by Alkett also incorporated Panzer III escape hatches on either side, a feature which is not present in other models. The Alkett 20 production run had same 10 speed transmission as Ausf.A, narrow 360mm tracks of Ausf.A, and incorporated the same return roller position as the Ausf.A.

The StuG III Ausf. B: (Sd.Kfz 142)

This variant entered production in June 1940 and by May 1941 when production finally halted over 300 had been produced. This variant featured widened tracks which were increased to 38cm. Also introduced at the time were two rubber tires on each road-wheel which were accordingly widened from 520 x 79mm to 520 x 95mm. As they shared a common diameter both types of road-wheels were interchangeable. The troublesome 10 speed transmission was reduced to a more robust six speed arrangement. A major design enhancement was the change introduced whereby the front-most return rollers were re-positioned further forward. This reduced vertical movements of tracks before they were fed to the drive wheel and significantly reduced the chance of a track being thrown. In the middle of production of Ausf. B., the original drive wheel with 8 round holes was changed to a new cast drive wheel with 6 slots. This new drive wheel could take either 380mm tracks or 400mm tracks. 380mm tracks were not exclusive to new drive wheels.

StuG III Ausf. C: (Sd.Kfz 142)

This model entered production in April 1941 and 50 were produced. The main alteration was the gunner's forward view port above the driver's visor which was considered to be a shot trap and was discarded. The superstructure top was given an opening for the gunner's periscope. The Idler wheel was also redesigned to help avoid track shedding.

StuG III Ausf. D: (Sd.Kfz 142)

This was in production from May until September 1941 and 150 machines were produced. This was essentially a contract extension based on the Ausf. C. However an on-board intercom was installed, otherwise this vehicle was identical to the Ausf. C.

StuG III Ausf. E: (Sd.Kfz 142)

This variant was in production from September 1941 until February 1942 during which time 284 machines were produced. To the superstructure sides were added extended rectangular armoured

A demonstration of the building of the Sturmgeschütze and Panzer IIIs which appeared in the pages of Signal magazine.

boxes for radio equipment. Increased space allowed room for the stowage of six additional rounds ammunition maximum to be raised to 50. The other crucial addition was a close defence machine gun. One MG 34 and seven drum-type magazines were carried in the right rear side of the fighting compartment to protect the vehicle from enemy infantry and tank hunting teams. Vehicle commanders were officially provided with SF14Z stereoscopic scissor periscopes. Stereoscopic scissor type periscopes for artillery spotters may have been used by vehicle commanders from the start.

StuG III Ausf. F: (Sd.Kfz 142/1)

This entered production in March 1942 and the run continued until September 1942 during which time 366 machines were produced. This was the first real up-gunning of the StuG, and this variant was provided with the longer and more powerful 7.5 cm StuK 40 L/43 gun. Firing armour-piercing Panzergranat-Patrone 39, the Stuk 40 L/43 could penetrate 91mm of armor inclined 30 degrees from vertical at 500m, 82mm at 1000m, 72mm at 1500m, 63mm at 2000m. The addition of this main gun allowed the Ausf. F to engage most Soviet tanks at normal combat ranges. This change marked the phase in which the StuG was deployed as a tank destroyer rather than an infantry support vehicle. An exhaust fan was added to the rooftop to excavate fumes from spent shells, to enable firing of

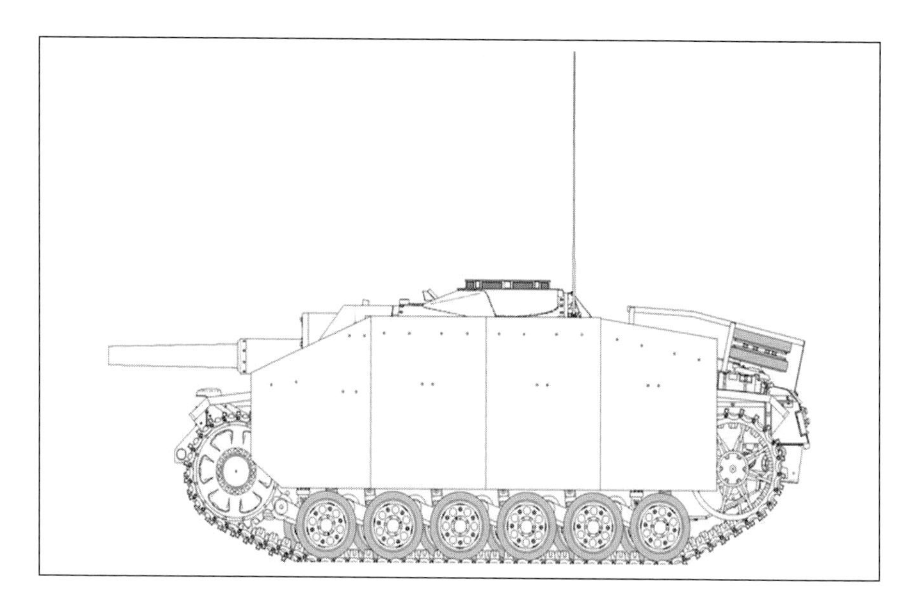

continuous shots. Additional 30mm armor plates were welded to the 50mm frontal armor from June 1942, making frontal armors 80mm thick. From June 1942, Ausf. F were mounted with approximately 13 inch (334mm to be exact) longer 7.5 cm StuK 40 L/48 gun. Firing the above mentioned ammunition, the longer L48 could penetrate 96mm, 85mm, 74mm, 64mm respectively (30 degrees from vertical).

StuG III Ausf. F/8: (Sd.Kfz 142/1)

September-December 1942, 250 produced. Introduction of an improved hull design similar to that used for the Panzer III Ausf. J / L with increased rear armor. This was the eighth version of Panzer III hulls, thus the designation "F/8." This hull has towing hook holes extending from the side walls. From October 1942, 30mm additional armors were bolted on to speed up the production line. From F/8, 7.5 cm StuK 40 L/48 gun becomes standard until the very last of the Ausf. G. Due to the lack of double baffle muzzle brakes, few L48 guns mounted on F/8 were fitted with the single baffle ball type muzzle brakes found in Panzer IV Ausf. F2.

StuG III Ausf. G (Sd.Kfz. 142/1;

December 1942– April 1945, 7,720 produced, 173 converted from Pz.Kpfw. III chassis): The final and by far the most common of the StuG series. The Ausf. G used the hull of the Pz.Kpfw. III Ausf. M. The upper superstructure was widened: welded boxes on either sides were abandoned. This new superstructure design increased its height to 2160mm. The backside wall of the fighting compartment got straightened, and the ventilation fan on top of the superstructure was relocated to the back of the fighting compartment. From March 1943, the driver's periscope was abandoned. From May 1943, side hull skirts (schurzen) were fitted to G models for added armor protection particularly against anti-tank rifles. Side skirts were retro-fitted to some Ausf. F/8 models, as they were be fitted to all front line StuGs and other tanks by June 1943 in preparation for the battle of Kursk. Mountings for side skirts proved inadequate, and many were lost in the field. From March 1944, improved mounting was introduced, and as a result side skirts were seen more often with late model Ausf G. From May 1943, 80mm thick plates were used for frontal armor instead of two plates of 50mm+30mm. However, there was a backlog of completed 50mm armor models. For those, 30mm of additional armor still had to be welded or bolted on, until October 1943.

A rotating cupola with periscopes was added for the commander for Ausf G. However, from September 1943, a lack of ball bearings (resulting from bombing of Schweinfurt–Regensburg mission) forced cupolas to be welded on. Ball bearings were once again installed from August 1944. Shot deflectors for cupolas were first installed from October 1943 from one factory, to be installed on all StuGs from February 1944. Some vehicles without shot deflectors carried several track pieces wired around the cupola for added protection.

Later G versions from November 1943, were fitted with the Topfblende (pot mantlet), often called Saukopf (pig's head), gun mantlet without coaxial mount. This cast mantlet with an organic shape was more effective at deflecting shots than the original box mantlet armor of varying thickness between 45mm and 50mm. Lack of large castings meant that the box mantlet was also produced until the very end. A coaxial machine gun was added first to box box mantlets from June 1944, and then to cast Topfblende from October 1944, in the middle of Topfblende mantlet production. With an addition of coaxial, all StuGs carried two MG 34 machine guns from fall of 1944. Some previously completed StuGs with box box box mantlet had a coaxial machine gun hole drilled to retrofit a coaxial machine gun, while Topfblende produced from November 1943 to October 1944 without machine gun opening could not be tampered with. Also from November 1943, all metal return rollers of a few different types were used due to lack of rubber supply. Zimmerit anti-magnetic coating to protect vehicles from magnetic mines was used from September 1943 to September 1944 only.

FURTHER VARIANTS

Sturmhaubitze III, Variant G

In 1942, a variant of the StuG III Ausf. F was designed with a 105 mm (4.1 in) howitzer instead of the 7.5 cm StuK 40 L/43 cannon. These new vehicles, designated StuH 42 (Sturmhaubitze 42, Sd.Kfz 142/2), were designed to provide infantry support with the increased number of StuG III Ausf. F/8 and Ausf. Gs being used in the anti-tank role. The StuH 42 mounted a variant of the 10.5 cm leFH 18 howitzer, modified to be electrically fired and fitted with a muzzle brake. Later models were built from the StuG III Ausf. G chassis as well as StuG III Ausf. F and Ausf. F/8 chassis. The muzzle brake was often omitted due to the scarcity of resources later in the war. 1,211 StuH 42 were produced from October 1942 to 1945.

In 1943, 10 StuG IIIs were converted to StuG III (Flamm) configuration by replacing the main gun with a Schwade flamethrower. These chassis were all refurbished at the depot level and were a variety of pre-Ausf. F models. There are no reports to indicate any of these were used in combat and all were returned to Ausf. G standard at depot level by 1944.

In late 1941 the StuG III chassis was selected to carry the 15 cm sIG 33 heavy infantry gun. These vehicles were known as Sturm-

Infanteriegeschütz 33B. Twenty-four were built of which twelve vehicles saw combat in the battle of Stalingrad where they were destroyed or captured. The remaining twelve vehicles were assigned to 23rd Panzer Division.

Some StuG III were also made from a Panzer III chassis but fitted with the bogie suspension system of the Panzer IV tank. Only about 20 were manufactured. The intention was to simplify field repairs but this did not work out as planned and the model was cancelled. Due to a dwindling supply of rubber, rubber saving road wheels were tested during 8th-14th November 1942, but did not see production.

Bombing raids on Alkett factory resulted in significant drops in StuG III production in November 1943. To make up for the loss of production, Krupp displayed a substitution StuG on a Panzer IV chassis to Hitler on 16th–17th December 1943. From January 1944, the StuG IV, based on the Panzer IV chassis and with a slightly modified StuG III superstructure, entered production.

Sturmgeschütze IV

The Sturmgeschütz IV resulted from Krupp's effort to supply an assault gun. As Krupp did not build Panzerkampfwagen IIIs, they used the Panzerkampfwagen IV chassis in combination with a slightly modified Sturmgeschütz III superstructure.

Initial Project: The first known proposal for a Sturmgeschütz on the Panzer IV chassis is in Krupp drawing number W1468 dated February 1943. This initial drawing unitized the outdated Sturmgeschütz Ausf. F superstructure on a Panzer IV chassis 9. This proposal had a sloped front superstructure with a combat weight of 28.26 tons. Krupp abandoned it in February 1943 because it was too heavy. Plans for the StuG IV were halted.

Another Project: During the Führer Conference of August 19th–22nd, 1943, after the battle of Kursk, Hitler had seen reports of the StuG III out-performing the Panzer IV within certain restraints of how they were deployed. Convinced that a tank-hunter version would be superior to the tank version, Hitler planned to switch Panzer IV production to "Panzerjäger IV" production as soon as possible. It was to mount the same 7.5 cm L70 used for the Panther. Another manufacturer, Vomag built a prototype Panzerjager IV with 7.5 cm L/48 gun and demonstrated it on October 20th, 1943. It was later re-designated as Jagdpanzer IV Ausf. F. As the Jagdpanzer IV was already being produced by Vomag, the StuG IV may not have materialized had it not been for the major disruption of StuG III production, and the scarce supply of the 7.5 cm L/70 gun designated for the Jagdpanzer IV.

Restart of the StuG IV: In November 1943, Alkett, a major StuG III manufacturer, was bombed. Alkett produced 255 StuG III in October 1943, but in December this fell to just 24 vehicles. On December 6th–7th, 1943, at a conference with Hitler, he welcomed the suggestion of taking the StuG III superstructure and mounting it on a Panzer IV chassis. The StuG IV could be more quickly manufactured than the Jagdpanzer IV at the time. This re-started the Sturmgeschütz IV project. This time, the superstructure of the StuG III Ausf. G was mounted on a Panzer IV chassis 7, with a box compartment for the driver added. Combat weight was 23,000kg, lighter than the 23,900kg for the StuG III Ausf. G. On December 16th-17th, 1943, Hitler was shown the StuG IV, and approved it. To make up for the large deficit in StuG III production, StuG IV production received full support.

THE CONTEMPORARY VIEW NO.9

NEW 75-MM GERMAN ASSAULT GUN

Extracted from Tactical and Technical Trends, No. 51, October 1944

A late model German 75-mm assault gun, only known model of this type of weapon mounted on the Pz. Kpfw. IV chassis, has been examined by Allied ordnance experts. It is designated as the 7.5-cm Stu. K. 40 (L/48).

Of 48 calibers in length, the gun has the same performance as the 7.5-cm Pak 40. The motor carriage. with the same performance as the Pz. Kpfw. IV, is designated Stu. G. IV für 7.5-cm. Stu. K. 40 (L/48).

This gun is characterized by heavy frontal armor and a low silhouette, typical of assault guns in general. The mantlet is cast in one piece, contrary to the usual German practice of welding. It covers the recuperator and is rounded instead of being in the familiar keystone shape. The brackets on the sides, looking like oversize fishhooks, are for attaching skirting armor. The wrinkled-canvas appearance of the sides of the vehicle is given by Zimmerit, a composition intended to prevent adhesion of magnetic mines.

From December 1943 to May 1945, Krupp built 1,108 StuG IVs and converted an additional 31 from battle-damaged Panzer IV hulls. While the number is smaller than the 9,000+ StuG III, the StuG IV supplemented and fought along with StuG III during 1944-45, when they were most needed.

The StuG IV became known as an effective tank killer, especially on the Eastern Front.

It had a four-man crew, and was issued mainly to infantry divisions.

- Commander in hull left rear
- Gunner in hull left center
- Loader in hull right rear
- Driver in hull left front

Field modifications were made to increase the vehicle's survivability, resulting in diversity to already numerous variants: cement plastered on the front superstructure, older Ausf.C/D retrofitted with a Kwk40 L48 gun, Ausf.G mounting Panzer IV cupola, a coaxial MG34 through a hole drilled through the front mantlet.

THE CONTEMPORARY VIEW NO.10

AFTER-ACTION REPORT

An after-action report from Panzerjager-Kompanie 1045 with StuG III in the Nachrichtenblatt der Panzertruppen, December 1944

"The company was prepared as divisional reserves. The enemy attacked one morning after a half-hour pummeling of artillery preparatory fire with heavy air support, and about 30 T-34 tanks and mechanized infantry deployed on a wide front. The enemy tried to force a breakthrough with portions of 5 or 6 divisions. The terrain was unusually favorable for the enemy. Above all, the forested areas provided him with suitable firing positions and assembly areas.

The company went into action with 9 Sturmgeschütz, and on the first day was able to knock-out or destroy the following within three hours:

16 T34, 2 mortars, 1 KVI, 2 observation points with radio, 2 T-34 (immobile), 1 anti-tank gun, 17 machine guns, 1 infantry gun

On the second day:

2 T34, 3 anti-tank weapons, 1 self-propelled gun, 2 grenade launchers, 21 machine guns, 2 anti-tank guns

The tanks were knocked-out at ranges of 600-800 meters. In a period of 15 minutes, one StuG was able to hit five tanks out of a column. The enemy didn't fire a single aimed round. The remaining T-34 tanks were individually hunted down. One T-34 was knocked out at a range of 1000 meters with 3 rounds."

A StuG III destroyed in Normandy, 1944. This vehicle appears to have suffered a catastrophic internal explosion.

CHAPTER 5

THE MAIN CHARACTERISTICS

This development of the Sturmgeschütz mounts the Stu. K. 40 assault gun, with muzzle brake, on a Pz. Kpfw. III chassis. The gun is identical to the Kw.K. 40 with the exception that in the Stu. K. 40 the buffer and recuperator systems are mounted on each side of the barrel to accommodate the S. P. mounting, while the Kw.K. 40 mounts the recoil mechanism above the barrel. The road performance of this vehicle approached that of the Pz. Kpfw. III tank.

The superstructure is a box-like arrangement, closed in on the top and welded to the chassis. A commander's cupola, loader's entrance hatch, and an opening for the sight are provided on the roof of the superstructure. There is no hull entrance door. A circular hole approximately 9 inches in diameter, cut in the rear of the superstructure, houses a small electric fan which draws air into the fighting compartment. A circular plate 12 inches in diameter and 30mm thick is fitted over the hole at a distance of 3 inches from the superstructure by means of three bolts. The thickness of superstructure armor is as follows: front vertical plate 50 + 30mm, gun mantlet 50mm, sides 30mm, top 20mm. Side skirting armor is often mounted on this vehicle.

The gun has a length of 126 inches, without the muzzle brake, a muzzle velocity of 2525 f/s for the 15.0 lb. projectile and an effective range of 2,000 yards firing A.P.C. shell. Its approximate radius of elevation is -5° to + 20°; traverse 10° left and right. It utilizes the following types of ammunition—A.P.C., H.E., and hollow charge. The penetration of A.P.C. shell against homogeneous plate is reported as follows: 500 yards, 4.0 inches at 30° obliquity; 4.8 inches normal—1,000 yards, 3.6 inches at 30°, 4.3 inches normal—2,000 yards, 2.8 inches at 30°, 3.4 inches normal.

SPECIFICATIONS

Weight	26.5 tons
Length	17 ft., 9 ins.
Width	9 ft., 7 ins.
Height	6 ft., 5 ins.
Ground clearance	14 ins.
Tread centers	8 ft., 2 1/2 ins.
Ground contact	9 ft., 4 1/2 ins.
Width of track	15 ins.
Pitch of track	43 ins.
Track links	90
Fording depth	3 ft.
Theoretical radius of action	
Roads	100 miles
Cross-country	60 miles
Speed	
Roads	25 m.p.h.
Cross-country	15 m.p.h.
Armor	
Front plate	50 + 30 mm
Sides	30 mm
Armament	7.5 cm Stu. K. 40
	2 Machine pistols
Ammunition	54 rds.
Engine	Maybach, HL 120 TRM, V-12, 320 hp.
Transmission	6 speeds forward, 1 reverse
Steering	Epicyclic, clutch brake
Crew	4

STURMGESCHÜTZ LG. 7.5 CM STU. K. (SD. KFZ. 142): S.P. ASSAULT GUN

This version of the Sturmgeschütz is similar to its predecessor, the Stu. G. 7.5 cm K., except for its armament. The 7.5 cm Kw.K. short gun which appeared in the original Sturmgeschütz was replaced by the 7.5 cm Stu. G. lg. K., 87 inches long as shown above. The arrangement of the recoil mechanism also differs, the long version having what appears to be the original mounting provided for the Kw.K. 40 and is evidently a forerunner in its design and development. Photographs show the 7.5 cm Stu. G. lg. K. to be equipped at times with a muzzle brake. The front superstructure has been modified to accommodate the improved armament.

This vehicle is made up of the Pz. Kpfw. III tank chassis and an early development of the 7.5 cm Pak 40, and includes the hull and superstructure improvements made at this time on the original Sturmgeschütz. A roof was added, a commander's cupola, smoke pot projectors and heavier front plate. The piece represents the trend in design from the short barrel 7.5 cm Kw.K. to the Stu. K. 40.

SPECIFICATIONS (Sturmgeschütz LG 7.5 cm Stu. K.)

Weight	21 tons
Length	17 ft., 9 ins.
Width	9 ft., 7 ins.
Height	6 ft., 5 ins.
Ground clearance	15 ins.
Tread centers	8 ft., 2 1/2 ins.
Ground contact	9 ft., 4 1/2 ins.
Width of track	15 ins.
Pitch of track	4 3/4 ins.
Track links	90
Fording depth	3 ft.
Theoretical radius of action	
Roads	100 miles
Cross-country	60 miles
Speed	
Roads	25 m.p.h.
Cross-country	15 m.p.h.
Armor	
Front plate	50 mm
Sides	30 mm
Armament	7.5 cm lg. Kw.K.
Ammunition	84 rds.
Engine	Maybach HL 120 TRM, 320 hp.
Transmission	6 speeds forward, 1 reverse
Steering	Epicyclic clutch brake
Crew	4

STURMGESCHÜTZ 7.5 CM K. (SD. KFZ. 142): S.P. ASSAULT GUN

The Sturmgeschütz is an assault weapon. Unlike the two other classes of self-propelled guns, antitank and artillery, which consist merely of guns placed in the hull of a tank with shields erected around the front and sides thereof, the assault gun is built into the hull and is consequently nearer the ground and has a much more solid superstructure built round the gun. The original Sturmgeschütz consisted of the turretless chassis of a Pz. Kpfw. III tank, upon which was mounted the Stu. G. 7.5 cm K., a short-barrelled (69.5 inch) piece found in the first models of the Pz. Kpfw. IV. Since the power plant and other mechanical components of the chassis of the Sturmgeschütz are identical to those of the Pz. Kpfw. III tank, and their weights are approximately the same, the performance data of the two are comparable.

The turret of the original tank has been removed and replaced by a squat superstructure, reducing the height of the vehicle from 8 feet, 3 inches as a tank to 6 feet, 5 inches as an assault weapon. The gun compartment is roofed over, but there is no rotating turret. The fighting compartment is armored as follows: front 53 mm, sides 43 mm, top 11 mm.

SPECIFICATIONS (Sturmgeschütz 7.5 cm K.)

Weight	21 tons
Length	17 ft., 9 ins.
Width	9 ft., 7 ins.
Height	6 ft., 5 ins.
Ground clearance	14 ins.
Tread centers	8 ft., 2 1/2 ins.
Ground contact	9 ft., 4 1/2 ins.
Width of track	15 ins.
Pitch of track	4 3/4 ins.
Track links	90
Fording depth	3 ft.
Theoretical radius of action	
Roads	100 miles
Cross-country	60 miles
Speed	
Roads	28 m.p.h.
Cross country	15 m.p.h.
Armor	
Front plate	50 mm
Sides	30 mm
Armament	7.5 cm Kw.K.
Ammunition	84 rds.
Engine	Maybach V-12, HL 120 TRM, 320 hp.
Transmission	6 speeds forward, 1 reverse
Steering	Epicyclic clutch brake
Crew	4

The gun, which is mounted low in the hull and fires forward, is identical to the 7.5 cm Kpfw. K. short tank gun, originally the main armament of the Pz. Kpfw. IV tank. It is primarily a close support weapon, the ammunition scale comprising only 25% A.P. against 10% smoke and 65% H.E.; its armor-piercing performance is relatively poor. Its muzzle velocity and maximum range firing H.E. shell is 1,378 f.s. and 6,758 yards, respectively. The penetration of A.P.C.B.C. shell against homogeneous armor is reported as follows: 500 yards, 1.81 inches at 30° obliquity, 2.16 inches normal—1000 yards, 1.61 inches at 30°, 1.97 inches normal—1,200 yards, 1.57 inches at 30°, 1.89 inches normal.

During the present war the Germans have been placing increasing emphasis on the class of weapon known as assault artillery. Consisting of assault guns and howitzers, assault artillery pieces should not be confused with other types of self-propelled guns, for each of the four types of German self-propelled guns has definite characteristics, and each follows prescribed tactics peculiar to its type. In contrast to assault artillery, self-propelled artillery provides indirect fire in the normal field artillery fashion, and is protected by open-top armored shields proof against only small-arms fire and shell fragments. Tank destroyers, which are armored like self-propelled artillery, are used by companies in counterattacking tank breakthroughs, each platoon concentrating on a single tank. Self-propelled infantry howitzers are also armored like self-propelled artillery; they are the infantry howitzers of armored infantry. Recent tank destroyers like the Jagdpanther have assault gun characteristics. However, Jagdpanthers are unique in that they fire only on long range targets.

German assault guns, like tanks, are often fitted with 5-mm side armor plates designed to explode hollow-charge (bazooka) projectiles before they hit the main armor. These plates bolt on in sections on frames. On the march, they are stacked on the rear.

The most common type of assault gun is the 7.5-cm Sturmgeschütz 40 (above). It supersedes the original 75-mm assault gun 7.5-cm Sturmgeschütz, shown on the next page. The chief difference between the original and the present version is the high-velocity 75. The long 75 permits assault guns to combat tanks and also affords greater accuracy at longer ranges. Assault howitzers are designed after the same principles as assault guns. In fact, the 105-mm 10-cm Sturmhaubitze 42 looks just like the Stu.G. 40 above. Somewhat bigger is the 150-mm 15-cm Sturmpanzer 43 ("Grizzly Bear"). Assault howitzers are used in the same way as assault guns. The Nazis believe that their large shells have great effect on morale.

Assault guns are reserved for attacks and counterattacks. Their low silhouette permits them to move close behind the leading infantrymen, and yet retain a maximum of security against hostile observation and hits from hostile weapons. The first type of assault gun, which mounts a short 75-mm gun, is shown here.

THE CONTEMPORARY VIEW NO.II

POWERED ARTILLERY

Extracted from Recognition Journal, September 1944

German medium assault gun is a thick-barreled 105-mm. gun mount on the chassis of the PzKw III tank. This weapon looks almost exactly like the 75-mm. Sturmgeschütz (overleaf).

Many German self-propelled guns are modified captured material.

German self-propelled guns have increased in importance with the Wehrmacht's withdrawal on all fronts. Retreating German armored divisions have left behind hundreds of these mobile assault and antitank guns.

Last December the Journal published pictures of the principal German self-propelled guns. These and many

This captured Czech PzKw 38 chassis that mounts a 75-mm. is a favorite German self-propelled weapon. The gun is housed in a tall, open-topped, slant-sided shield well to the rear.

To pierce the Gustav Line, the Allied troops had to beat their way past many 75-mm. self-propelled assault guns on PzKw III chassis. These last-ditch defenders made the going tough for the Allies. A squat forward superstructure has replaced the PzKw III turret. It is closed in on top and welded to the chassis.

More familiar version of German 75-mm. gun on the Czech chassis has cone-shaped gun housing set well forward. Czech suspension has short wheel base of four large bogies

others have since confronted the Allies in the U.S.S.R., Italy and France. Some of the guns that we captured in Tunisia have continued to appear in large numbers in Europe. They include the 75-mm. and 105-mm. assault guns on the PzKw III chassis, the 75-mm. on the Czech PzKw 38 chassis and on the French Lorraine chassis.

Soviet troops first encountered the self-propelled 88-mm. gun, Ferdinand, in 1943; U.S. and British troops met it in Italy. How they dealt with the 72-ton monster is shown on page 14. Ferdinand remains the only German self-propelled gun that was made to order from start to finish -- an original design. The others have all been adaptations.

The newest German SP guns continue to be adaptations of present equipment. They illustrate a trend toward greater firepower on PzKw IV and VI chassis.

A StuG III D of z.v.B 288

7,5 CM STURMKANONE (STUK) 37 L/24

Penetration Table

Round	Type	Warhead Weight	Muzzle Velocity	Penetration at 30°			
				100 m	500 m	1,000 m	1,500 m
	HE	5.7 kg	420 m/s	-	-	-	-
K.Gr.rot Pz.	-	6.8 kg	385 m/s	41 mm	39 mm	35 mm	33 mm
Gr.38 HL	HEAT	4.5 kg	452 m/s	45 mm	45 mm	45 mm	45 mm
Gr.38 HL/A	HEAT	4.4 kg	450 m/s	70 mm	70 mm	70 mm	70 mm
Gr.38 HL/B	HEAT	4.75 kg	450 m/s	75 mm	75 mm	75 mm	75 mm

Accuracy Table

Round		Range			
		100 m	500 m	1,000 m	1,500 m
K.Gr.rot Pz.	Training	100%	100%	98%	74%
	Action	100%	100%	73%	38%
Gr.38 HL	Training	100%	100%	92%	61%
	Action	100%	99%	60%	26%

7,5 CM STURMKANONE (STUK) 40 L/43

Penetration Table

Round	Type	Warhead Weight	Muzzle Velocity	Penetration at 30°				
				100 m	500 m	1,000 m	1,500 m	2,000 m
	HE	5.7 kg	550 m/s	-	-	-	-	
Pzgr.39	APCBC	6.8 kg	750 m/s	94 mm	86 mm	77mm	67 mm	58 mm
Pzgr.40	APCR	4.1 kg	920 m/s	126 mm	108 mm	87 mm	69 mm	-
Gr.38 HL/C	HEAT	5 kg	450 m/s	100 mm	100 mm	100 mm	100 mm	100 mm

Accuracy Table

Round		Range						
		100 m	500 m	1,000 m	1,500 m	2,000 m	2,500 m	3,000 m
Pzgr.39	Training	100%	100%	99%	77%	48%	30%	17%
	Action	100%	99%	71%	33%	15%	8%	4%
Pzgr.40	Training	100%	100%	95%	66%	21%	-	-
	Action	100%	98%	58%	24%	6%	-	-
Gr.38 HL/C	Training	100%	100%	85%	42%	20%	-	-
	Action	100%	100%	45%	15%	6%	-	-

7,5 CM STURMKANONE (STUK) 40 L/48

Penetration Table

Round	Type	Warhead Weight	Muzzle Velocity	Penetration at 30°				
				100 m	500 m	1,000 m	1,500 m	2,000 m
	HE	5.7 kg	550 m/s	-	-	-	-	-
Pzgr.39	APCBC	6.8 kg	750 m/s	106 mm	96 mm	85 mm	74 mm	64 mm
Pzgr.40	APCR	4.1 kg	930 m/s	143 mm	120 mm	97 mm	77 mm	-
Gr.38 HL/C	HEAT	5 kg	450 m/s	100 mm	100 mm	100 mm	100 mm	100 mm

Accuracy Table

Grenade		Range						
		100 m	500 m	1,000 m	1,500 m	2,000 m	2,500 m	3,000 m
Pzgr.39	Training	100%	100%	99%	77%	48%	30%	17%
	Action	100%	99%	71%	33%	15%	8%	4%
Pzgr.40	Training	100%	100%	95%	66%	21%	-	-
	Action	100%	98%	58%	24%	6%	-	-
Gr.38 HL/C	Training	100%	100%	85%	42%	20%	-	-
	Action	100%	100%	45%	15%	6%	-	-

10,5 CM STURMHAUBITZE (STUH) 42 L/28

Penetration Table

Round	Type	Warhead Weight	Muzzle Velocity	Penetration at 30°				
				100 m	500 m	1,000 m	1,500 m	2,000 m
	HE	14.8 kg	470 m/s	–	–	–	–	–
	APCBC	14.0 kg	750 m/s	65 mm	59 mm	54 mm	50 mm	46 mm

POLAND 1939

No Sturmgeschütz units were available for the Polish Campaign.

FRANCE 1940

The first 24 StuG III Ausf A's produced equipped Sturmartillerie Batteries 640, 659, 660 and 665. Each battery had six assault guns in three platoons (with two assault guns each K.St.N.445). Due to production delays, Sturmartillerie Batterie 640 received four Sd.Kfz. 251s instead of the Sd.Kfz. 253s. For the same reason Sturmartillerie Batterie 660 received "turretless" Pz.Kpfw. I munitions carriers instead of their allotted Sd.Kfz. 252s.

Sturmartillerie Batterie 640 became organic to Infantry Regiment Grossdeutschland and was renamed the 16. Sturmbatterie/Infantrie Regiment “Grossdeutschland”

Six assault guns were issued to SS-Sturmartillerie battery of Leibstandarte SS Adolf Hitler division.

Two more batteries were formed - 666th and 667th but didn't see combat in France.

THE BALKANS AND GREECE 1941

The Sturmgeschütz-Abteilung 184, 190 & 191 and the 16. Sturmbatterie /Infantrie Regiment “Grossdeutschland” were employed in the campaign in Greece and Yugoslavia.

Each Abteilung consisted of three Batterie of six Sturmgeschütz each; for a total of 18.

NORTH AFRICA

5.Panzerjägerkompanie/Sonderverband z.b.V 288

Three Ausf D vehicles were sent to North Africa and saw service with Sonderverband z.b.V 288 - special deployment unit.

Sonderverband 288 was raised on July 1st, 1941 at Potsdam (near Berlin). It was composed of units from all over Germany. It originally contained crack units from all branches of the German army: anti-tank men, Alpine troops, engineers, and eventually 3 of the new StuG.III Ausf D assault guns. There was also a company of Brandenburgers, trained for special operations behind enemy lines.

These units were intended to prepare the way for the DAK as it flowed across the Nile and into the Middle East and on to India. Their most important component was a group of interpreters with their own printing presses. These men knew all the languages that would be needed, from Arabic and Persian dialects to Hindi, Urdi and Sanskrit. A group of forgers and engravers with printing presses for counterfeiting foreign currency were part of the unit. There was also a specialized group who had been trained to seize and rebuild the oil fields of the region.

When it became obvious that Rommel’s rush to the Nile was not

A StuG III D of z.v.B 288

going as planned, the 288th minus its interpreters and technicians was sent to Africa as special reinforcements.

During the Gazala Line battles the 288th saw action supporting the Italian Ariete against the Free French forces defending Bir Hacheim at the southern tip of the British defenses.

They also participated in the El Alamein battles, where the last of their StuG.III's were left behind and captured by the British.

TUNISIA 1942

Sturmgeschütz-Batterie 90 of the 10.Panzer Division

Four Sturmgeschütz, Ausf. F/8 were landed in Tunisia in November 1942.

LUFTWAFFE STURMGESCHÜTZ UNITS (INDEPENDENT)

Sturmgeschütz-Abteilung 1 der Luftwaffe
Sturmgeschütz-Brigade 1 der Luftwaffe (Apr 44)
Fallschirm-Sturmgeschütz-Brigade 11 (Jun 44)
Fallschirm-Sturmgeschütz-Brigade 111 (28 Mar 45)

- Jan 44 - formed with four batteries for I Fallschirm-Korps - 37 Semovente M/42 enroute to equip this unit in December 43.
- Mar 44 - 22 assault guns operational
- Sep 44 - entered combat vicinity of Nancy and destroyed
- Oct-Nov 44 rebuilt
- Dec 44 - 27 assault guns; attached to 5. Fallschirmjäger-Division, 7th Army (Ardennes)
- Mar 45 – With 5. Fallschirmjäger-Division at Nieder-Breisig

Sturmgeschütz-Abteilung 2 der Luftwaffe
Sturmgeschütz-Brigade 2 der Luftwaffe (Mar 44)
Fallschirm-Sturmgeschütz-Brigade 12 (Jun 44)
Fallschirm-Sturmgeschütz-Brigade 121 (28 Mar 45)

Mar 44 - formed with four batteries for II Fallschirm-Korps

- Jun 44 – committed to battle in Normandy
- Jun 44 - 11 combat-ready StuGs on hand
- Jul 44 - StuG III's and 3 StuH 42s serviceable
- Aug 44 - virtually wiped out during the Falaise battles; one StuH 42 escapes
- Sep 44 - rebuilding near Köln-Wahn with 5 assault guns
- Sep 44 - transferred to the Arnhem area to provide support for 7. Fallschirmjäger-Division
- Jan 45 - fully equipped with assault guns while stationed at Amersfoot
- Feb 45 - attached to 7. Fallschirmjäger-Division during the Reichswald battle

Fallschirm-Sturmgeschütz-Brigade Schmitz
Fallschirm-Sturmgeschütz-Brigade 21 (1 Jan 45)

Fallschirm-Sturmgeschütz-Brigade 210 (28 Mar 45)

- Jan 45 - in Italy with four batteries
- Feb 45 - I Parachute Corps, 10th Army, Army Group Ligurien under the title of Fallschirm-Sturmgeschütz-Brigade

USING K.ST.N. 1159 – (14 GESCH.) (JUNE 1943)

3. Panzergrenadier Division

Panzer-Abteilung 103 (May 1943)
1. Kompanie - (14x StuG III 7.5cm)
2. Kompanie - (14x StuG III 7.5cm)
3. Kompanie - (14x StuG III 7.5cm)

10. Panzergrenadier Division

Panzer-Abteilung 7 (Sep 1943)
1. Kompanie - (14x StuG III 7.5cm)
2. Kompanie - (14x StuG III 7.5cm)
3. Kompanie - (14x StuG III 7.5cm)

16. Panzer Division

2. Panzer Regiment (Jun 1943)
9. Kompanie - (14x StuG III 7.5cm)
10. Kompanie - (14x StuG III 7.5cm)
11. Kompanie - (14x StuG III 7.5cm)

16. Panzergrenadier Division

Panzerjäger-Abteilung 228 (Sep 1943)
1. Kompanie - (14x StuG III 7.5cm)

18. Panzergrenadier Division

Panzer-Abteilung 118 (Sep 1943)
1. Kompanie - (14x StuG III 7.5cm)
2. Kompanie - (14x StuG III 7.5cm)
3. Kompanie - (14x StuG III 7.5cm)

20. Panzergrenadier Division

Panzer-Abteilung 8 (Sep 1943)
1. Kompanie - (14x StuG III 7.5cm)

2. Kompanie - (14x StuG III 7.5cm)
3. Kompanie - (14x StuG III 7.5cm)

25. Panzergrenadier Division
Panzer-Abteilung 5 (Sep 1943)
1. Kompanie - (14x StuG III 7.5cm)
2. Kompanie - (14x StuG III 7.5cm)
3. Kompanie - (14x StuG III 7.5cm)

29. Panzergrenadier Division
Panzer-Abteilung 129 (May 1943)
1. Kompanie - (14x StuG III 7.5cm)
2. Kompanie - (14x StuG III 7.5cm)
3. Kompanie - (14x StuG III 7.5cm)

60. Panzergrenadier-Division
(Panzergrenadier-Division Feldherrnhalle)
Panzer-Abteilung Feldherrnhalle (Aug 1943)
1. Kompanie - (14x StuG III 7.5cm)
2. Kompanie - (14x StuG III 7.5cm)
3. Kompanie - (14x StuG III 7.5cm)

90. Panzergrenadier Division
Panzer-Abteilung 190 (Feb 1944)
1. Kompanie - (14x StuG IV 7.5cm)
2. Kompanie - (14x StuG IV 7.5cm)
3. Kompanie - (14x StuG IV 7.5cm)

USING K.ST.N. 1158 – (22 GESCH.) (JUNE 1943)

14. Panzer Division
Panzer Regiment 36 (Jul 1943)
10. Kompanie - (22x StuG III 7.5cm)
12. Kompanie - (22x StuG III 7.5cm)
(plus two Kompanien of Pz.Kw. IV)

24. Panzer Division
Panzer Regiment 24 (Jun 1943)

9. Kompanie - (22x StuG III 7.5cm)
11. Kompanie - (22x StuG III 7.5cm)
(plus two Kompanien of Pz.Kw. IV)

STURMGESCHÜTZ ISSUED IN PLACE OF PZ.KPFW. IV AND PANZER IV/70(V) & JAGDPANTHER.

519. Panzerjäger-Abteilung

Panzerjäger-Abteilung 519 (Sep 1944)

- Stabs - (4x StuG III 7.5cm)
- Kompanie - (22x StuG III 7.5cm)
- Kompanie - (22x StuG III 7.5cm)

559. Panzerjäger-Abteilung

Panzerjäger-Abteilung 559 (Aug 1944)

- Stabs - (4x StuG III 7.5cm)
- Kompanie - (22x StuG III 7.5cm)
- Kompanie - (22x StuG III 7.5cm)

111. Panzer-Brigade

Panzer-Abteilung 2111 (Sep 1944)
4. Panzerjäger Kompanie - (10x StuG III 7.5cm)

112. Panzer-Brigade

Panzer-Abteilung 2112 (Sep 1944)
4. Panzerjäger Kompanie - (10x StuG III 7.5cm)

113. Panzer-Brigade

Panzer-Abteilung 2113 (Sep 1944)
4. Panzerjäger Kompanie - (10x StuG III 7.5cm)

2. Panzer Division

Panzer Regiment 3 (Nov 1944)
5. Kompanie - (14x StuG III 7.5cm)
6. Kompanie - (14x StuG III 7.5cm)
Panzerjäger-Abteilung 38 (Nov 1944)

- Stabs - (1x StuG III 7.5cm)
- Kompanie - (10x StuG III 7.5cm)
- Kompanie - (10x StuG III 7.5cm)

9. Panzer Division

I Batallion (Dec 1944)

- Panzer Regiment 33
- Kompanie - (14x StuG III 7.5cm)

116. Panzer Division

Panzer Regiment 33 (Dec 1944)

5. Kompanie - (14x StuG III 7.5cm)

510. Panzerjäger-Abteilung

Panzerjäger-Abteilung 510 (Feb 1945)

- Kompanie - (10x StuG III 7.5cm)
- Kompanie - (10x StuG III 7.5cm)
- Kompanie - (10x StuG III 7.5cm)

II Panzer Abteilung

Panzer Regiment 2 (Mar 1945)

- Stabs - (1x StuG III 7.5cm)
- Kompanie - (10x StuG III 7.5cm)
- Kompanie - (10x StuG III 7.5cm)
- Kompanie - (10x StuG III 7.5cm)

CHAPTER 6

THE STURMGESCHÜTZ AND OTHER NATIONS

Beginning in 1943, Hitler ordered that a small number of Sturmgeschütz III production be delivered to foreign countries. In most cases these were active allies of Germany (Spain being the exception). All of these allies signed separate armistices or changed sides by late 1944.

Bulgaria - Between February and December 1943 the Bulgarian armed forces received 55 Sturmgeschütz (most likely Ausf. G). These didn't see any action against the Soviets, but were rumored to have been used against the Wehrmacht when Bulgaria changed sides. The Bulgarians organized two Sturmgeschütz Abteilungen of 25 vehicles each.

The Bulgarian 1st Sturmgeschütz Abteilung was based in Sofia.

The Bulgarian 2nd Sturmgeschütz Abteilung in Plovdiv.

Finland - During the summer of 1943 the Finns received 30 Sturmgeschütz Ausf. G. A further 30 were shipped during Spring and early Summer of 1944. 15 were scheduled to be delivered in September of 1944, but were cancelled when Finland & Russia signed a cease fire on 9-Sept. In total the Finns fielded 59 StuG III's. Looking at the shipping inventory and the delivery numbers shows a discrepancy of one StuG. I have no evidence, but presume this was lost to accident or enemy action prior to the Finns taking delivery.

Hungary - From August to September 1944 the Hungarians took delivery of 40 Sturmgeschütz Ausf. G.

Italy - In May, 1943 the Italians took delivery of 5 Sturmgeschütz Ausf. G.

Spain - In October, 1943 the Spanish took delivery of 10 Sturmgeschütz Ausf. G.

Rumania - In the Winter of 1943 the Romanians took delivery of 6 Sturmgeschütz Ausf. G. From January to July of 1944 a further 94 units were delivered. Another 20 were scheduled in August, 1944 but due to the change in leadership (Coup d'Etat) the order was cancelled.

None of this Rumanian initial batch of 100 Stug IIIs survived the end of the war. 31 TAs were on the army inventory in November 1947. Most of them were probably StuG III Ausf. G and a small number of Panzer IV/70 (V), known as TAs T4. These TAs were supplied by the Red Army or were damaged units repaired by the Romanian Army. All German equipment was scrapped in 1954 due to the Army's decision to use Soviet armour.

StuG IIIs were also exported to other nations such as Bulgaria, Hungary, Italy, and Spain.

Many German Sturmgeschütz IIIs were stranded in Yugoslavia after the war. These were used by the Yugoslav Peoples Army until the 1950s.

After the Second World War the Soviet Union donated some of their captured German vehicles to Syria, which continued to use them along with other war surplus AFVs (like long-barreled Panzer IVs and T-34/85s) during the 1950s and up until the War over Water against Israel in the mid-1960s. By the time of the Six Days War all of them had been either destroyed, stripped for spare parts, or interred on the Golan Heights as static pillboxes.

The Soviet SU-76i self-propelled gun was based on captured StuG III and Panzer III vehicles. In total, Factory #37 in Sverdlovsk manufactured 181 SU-76i plus 20 commander SU-76i for Red Army service by adding an enclosed superstructure and the 76.2 mm S-1 tank gun.

More from the same series

Most books from the 'Hitler's War Machine' series are edited and endorsed by Emmy Award winning film maker and military historian Bob Carruthers, producer of Discovery Channel's Line of Fire and Weapons of War and BBC's Both Sides of the Line. Long experience and strong editorial control gives the military history enthusiast the ability to buy with confidence.

Tiger I in Combat | Tiger I Crew Manual | Panzers at War 1939-1942 | Panzers at War 1943-1945

Wolf Pack - the U boats | Poland 1939 | Luftwaffe Combat Reports | Sturmgeschütze

German Artillery in Combat | Panzer Combat Reports | The Panther V in Combat | German Tank Hunters

The Afrika Korps in Combat | Panzers I & II | Panzer III | Panzer IV

For more information visit www.pen-and-sword.co.uk